"The success of radio frequency identification will be through the convergence of supply chain execution applications and RFID technologies – a convergence that delivers true value in the supply chain – beyond basic compliance. With Printronix, companies can be confident in printing technology that complies with RFID standards and is designed to integrate with RFID-enabled solutions."

– Pete Sinisgalli
President and CEO
Manhattan Associates

"RFID for the supply chain is an ever moving target. This new edition hits the bulls eye! These updates are vital for anyone wanting to implement a successful, 'go forward' RFID solution."

– Greg Dixon
Chief Technology Officer
ScanSource, Inc.

"Within five years, RFID will become an essential tool for streamlining and securing global supply chains. Like any new technology, it is absolutely critical for companies starting out to make intelligent choices about implementation. This is the most comprehensive book out there on the subject."

– Vijay Sarathy
Director of Marketing and Strategy
Sun RFID Business Unit

"*RFID Labeling Second Edition* is a good reference for many of the new developments in the Supply Chain application of RFID technology. A clear, easy-to-read book."

– Rich Fletcher
Head of Packaging Special Interest Group
MIT

W9-BRM-640

"*RFID Labeling Second Edition* provides very useful information in an easy-to-read format. The insight from the case studies is invaluable. This book is a great educational resource for any organization considering RFID."

– Alok Ahuja
  Sr. Product Manager, RFID
  BPI Marketing
  Microsoft Corporation

"*RFID Labeling* will provide a solid foundation to those embarking on their own smart labeling projects."

– Dennis Vogel
  Manager, Product Marketing
  Advanced Technologies, Access BU
  Cisco Systems, Inc

"This book takes the complexity of implementation and brings it down to a common level. The graphics and pictures were very valuable. It is very easy to read, straight forward, and simple for the non-technical person."

– Doug Oathout
  Vice President of Office Solutions, Supplies and Maintenance
  IBM Printing Systems Division

"An excellent compendium on passive UHF RFID, and the forces that are accelerating its adoption in the retail consumer goods and military supply chains."

– Tom Pagel
  President, Label Products Division
  Nashua

RFID Labeling

Smart Labeling Concepts & Applications for the Consumer Packaged Goods Supply Chain

2ND EDITION

by Robert A. Kleist, Theodore A. Chapman, David A. Sakai, Brad S. Jarvis

PRINTRONIX®

Printronix, Inc.
14600 Myford Rd.
P.O. Box 19559
Irvine, CA 92623-9559

www.printronix.com

First edition, August 2004
Second edition, August 2005
Printed in the United States of America
Standard Book Number: ISBN 0-9760086-1-0
Library of Congress Catalog Control Number: 2005906997

Design: Eaton & Associates Design Company, Minneapolis

Contents

FOREWORD IX

DEFINITIONS XII

CHAPTER 1: INTRODUCTION 1

    REVOLUTIONARY TECHNOLOGY, OR COST BURDEN? 3

    Reducing Inventory Guesswork 4

    Why RFID Now? 5

    Using the Supply Chain for Competitive Advantage 7

    Much More than a Supply Chain Mandate 8

    BUILDING A BUSINESS CASE FOR RFID. 11

    Pushing RFID Down the Supply Chain 16

    Hype and Misconception 18

    PROTECTING INVESTMENTS DURING IMPLEMENTATION 20

    HOW THIS BOOK IS ORGANIZED 22

    Lessons Learned 23

CHAPTER 2: RFID BASICS 25

    RADIO WAVES 26

    European Frequency Regulatory Standard 30

    THE RFID TAG 32

    Tag Types 37

    Data Storage Capability 40

    Evolution of Tag Classes. 41

    Generation 1 Tag Specification. 43

    Generation 2 Tag Specification 45

    Tag Selection and Readability 49

PRINTER/ENCODERS 52

READERS 57

First Generation Basic Operation 59

Gen 2 Basic Operation 61

READER ANTENNAS 63

INTEROPERABILITY TESTING 69

THE GEN 2 STORY 70

CHAPTER 3: FROM UPC TO EPC 83

STANDARDS DEVELOPMENT THROUGH EPCGLOBAL 85

ELECTRONIC PRODUCT CODE 87

EPC Format 88

EPC Representations of Standard Identity Types 90

EPC & RFID COMPARED TO UPC AND BAR CODES 92

TAG SECURITY 100

Security Enhancements in Gen 2 Specification 102

ADDRESSING CONSUMER CONCERNS ABOUT RFID 106

CHAPTER 4: FROM BAR CODES TO SMART LABELS 109

ANATOMY OF A SMART LABEL 112

SELECTING THE RIGHT SMART LABEL FOR THE JOB 116

LABEL CERTIFICATION 119

ENCODING, PRINTING AND VALIDATING SMART LABELS 121

TROUBLESHOOTING TAG READING PROBLEMS 127

SMART LABELS COMPARED TO OTHER APPROACHES 128

IMPLEMENTING SMART LABELS 131

DEPENDENCIES AND POINTS OF CONTROL 131

ENTERPRISE-WIDE SMART LABEL PRINTING MANAGEMENT 134

CHAPTER 5: INDUSTRY INITIATIVES. 137

WAL-MART. 138

Wal-Mart RFID System Requirements 141

Advanced Shipping Notice Integration 142

Replenishment System Data Sharing. 143

OTHER RETAIL MANDATES. 146

European Retail 147

GLOBAL SUPPLY CHAINS. 151

RFID in China 151

Protecting the Global Food Supply 152

Challenges of Ocean Shipping 154

PHARMACEUTICAL INITIATIVES. 156

CHAPTER 6: DEPARTMENT OF DEFENSE INITIATIVE. 161

SENSE AND RESPOND LOGISTICS 163

RFID's Role in the DLA Transformation 165

GUIDELINES FOR IMPLEMENTATION 169

Updated Mil Standard for Package Labeling 171

Advanced Ship Notice 173

Gen 2 Transition. 174

Data Constructs 174

Supplier Funding for Implementation 176

CHAPTER 7: SMART START TO RFID-4 FRIENDLY PHASES 179

PHASE 1: GETTING STARTED 180

Establish the Team. 181

Feasibility Study. 182

Test Lab 186

Producing Labels for Testing 188

PHASE 2: TEST AND VALIDATION. 192

System Integration. 192

Validate Vendor Choices 193

Point-to-point Testing 194

PHASE 3: PILOT IMPLEMENTATION 194

PHASE 4: IMPLEMENTATION 198

VENDOR CHECKLIST 202

Certification Programs. 203

CHAPTER 8: CASE ANALYSIS 207

LAB TESTING. 209

LOCATION TESTING 210

LABEL PLACEMENT 211

READER ANTENNA PLACEMENT 214

Tools for Case Analysis 216

CHARACTERISTICS OF RF AFFECTING READ RATE. 219

CHAPTER 9: SMART LABELING APPROACHES 227

MANUAL "SLAP & SHIP". 231

SLAP & SHIP WITH EPC MANAGEMENT 234

AUTOMATED "SLAP & SHIP" 237

Meeting Compliance Mandates With Slap & Ship. 239

Smart Label Print & Apply Approaches 241

AUTOMATING MANUFACTURING PRODUCTION 244

SMART LABEL VALIDATION 245

A Pragmatic Path to Automation 246

CHAPTER 10: MANAGING EPC DATA COLLECTION 249

APPLICATION INFRASTRUCTURE. 250

Components of the RFID Network 252

Regulatory and Operational Requirements 253

Network Traffic 256

HOST TO READER COMMUNICATIONS 257

Gen 2 Host to Reader Interface 259

Application Level Events (ALE) Specification 261

Event Management 264

Reader Administration 265

Security and Quality of Service (QoS) 267

GLOBAL EXCHANGE OF SUPPLY CHAIN DATA 268

GLOBAL DATA SYNCHRONIZATION 271

CHAPTER 11: USING RFID DATA IN THE SUPPLY CHAIN 277

COMPLIANCE MANAGEMENT 279

Tracking Compliance Data 282

WAREHOUSE MANAGEMENT 288

RFID Data Used in Warehouse Operations 291

Yard Management 294

MANUFACTURING EXECUTION 297

Data Collection Integrated to ERP 298

OTHER SUPPLY CHAIN AREAS 299

Transportation Management 299

Customer Service 300

Management Dashboard 301

Event Management 302

Supplier Enablement 302

CHAPTER 12: A PARTNER'S APPROACH TO THE SUPPLY CHAIN 305

WAREHOUSE MANAGEMENT 306

DISTRIBUTED ORDER MANAGEMENT 308

TRADING PARTNER MANAGEMENT 310

TRANSPORTATION MANAGEMENT 311

RFID IN A BOX® 313

INTEGRATED LOGISTICS SOLUTIONS™ 315

CHAPTER 13: LESSONS LEARNED 317

FOOD INDUSTRY 318

Case 1: Chips in the Freezer 318

Case 2: Taking the Freeze out of RFID 324

Case 3: A New Generation Supply Chain for Snack Foods 330

ELECTRONICS INDUSTRY 338

Case 4: Smart Partnering for a High-tech Titan 338

MILITARY SUPPLIER 345

Case 5: Transportation Supplier Rides the RF Wave 345

PRINTRONIX PRODUCTS SECTION 351

SMART LABEL DEVELOPER'S KIT 352

SMART LABEL PRINTERS 354

ENCODE, PRINT AND APPLY RFID SMART LABEL APPLICATOR SYSTEM 358

RFID SMART LABELS 362

THERMAL BAR CODE PRINTERS – RFID READY 364

PrintNet ENTERPRISE PRINTER MANAGEMENT 368

OPTIONS AND ACCESSORIES 372

GENERAL PURPOSE INPUT AND OUTPUT MODULE 376

PROFESSIONAL SERVICES 378

PRINTRONIX CONTACT INFORMATION 380

INDEX 381

ABOUT THE AUTHORS 388

Foreword

Small in proportions, but monumental in its effects, RFID represents a big change in the retail industry. The recent initiatives of Wal-Mart, Target, the Department of Defense and other organizations, requiring case and pallet labeling for RFID, may seem trivial to the general public. That viewpoint is not shared by the relatively few people tasked with the job of bringing their companies into compliance over the next year or so. You may be affected already, perhaps as a member of your company's RFID task force, or as an employee of your company's supply chain.

RFID implementation is a real opportunity for those of us who support the retail industry. It is no small task, and we've observed a real lack of practical information. That is the reason for this handbook. Printronix is deeply involved in helping our clients meet RFID requirements. *RFID Labeling: Smart Labeling Concepts and Applications for the Consumer Packaged Goods Supply Chain* is part of our effort to help companies get started with the technology.

This book is a collaboration of information from our employee subject matter experts and business partners. The first five chapters cover the basics, with plenty of cross-references to later chapters that cover topics in more detail. We have made every attempt to provide accurate information, but we freely admit that neither we nor anyone else today are experts in all aspects of RFID. In fact, the speed of change in RFID technology for retail supply chains makes publishing a book on the subject a risky undertaking, and for this we ask your indulgence.

The subtitle of this book makes clear our point of view: smart labels offer the lowest cost, most practical, least disruptive way to implement RFID in the retail supply chain. With smart labels, you can succeed at achieving compliance with no business interruption. With smart labels, you can stream RFID into your current bar code labeling system, using proven tools to integrate both the physical tags and the associated EPC data, and even capture a database of information that provides traceability to each printed label.

Printronix has a rich 30-year history of enabling global printing solutions for supply chains. Companies worldwide use our thermal printers to produce bar code labels that guarantee 100 percent readable labels for retail packaged goods and government mandated identification requirements. We realize that printers are only an element of a package labeling process, so we collaborate with partners and industry groups to integrate seamless mission critical solutions. Our successful heritage includes many years of matching on-demand printing technologies to legacy applications and global enterprise management systems. We recognize that our solutions are only as good as the benefits that they bring to your supply chain, through increased efficiencies, the elimination of rework, and the availability of accurate and useful data that you can use to improve your business.

By attaching smart labels on cases and pallets, retail packaged good companies are placing themselves at the forefront of a new era. What may appear to be a forced change with no return on investments may actually become a way for companies to rethink and reengineer processes, enhance their value as business partners, and capture a profitable return.

As a member of EPCglobal, AIM Global Vendor Compliance Federation and the MIT RFID Packaging Special Interest Group, Printronix looks forward to helping you meet the goals of your RFID program. We would like to hear from you. Please take a moment to fill out and mail the reply card that you'll find inside the book.

August, 2005

Bob Kleist

This book is dedicated to Erwin Kelen, whose vision and inspiration made it a reality.

Acknowledgements: the authors wish to thank the following people for their significant contributions to this book: Carol Ballesty, Jose Basa, Scott Begbie, Bob Crum, Tim Eaton, Andy Edwards, David Fitzsimmons, Rick Fox, James Harkins, George Harwood, Karen Jensen, Van Le, Tim McGilloway, Jim McWilson, Andrew Moore, Steve Morris, Gayle Paride, and Toenya Rose.

Special thanks to Jerry Houston and Lisa Reickerd for their committment to getting this book done.

Definitions

Absorption
The degree to which materials change radio waves (electromagnetic energy) into current and heat.

Active tag
RFID tags having an active on-board transmitter, usually powered by battery, that constantly emits a signal, with a read range of 100 feet (30 m) or more. EPC Classes 3 and 4.

Adaptive frequency agility (AFA)
A characteristic of the new European air interface standard, EN 302 208, where readers listen first before broadcasting, thereby reducing collision with other transmissions.

AIM-USA
Automatic Identification Manufacturers, a USA trade association.

Air interface
The radio communications specification for a tag and reader, that describes the operating frequency (UHF 915 Hz), call and response characteristics (AM, half-duplex, pseudo-random frequency hopping, etc.), allowable transmission distances and applicable regulations.

Application Level Events (ALE)
Pending EPCglobal specification for reader management software.

Amplitude Modulation (AM)
Method of combining an information signal and an RF carrier, where a different voltage level is assigned to a digital 0 and 1.

Antenna
A radio frequency transducer. A receiving antenna converts an electromagnetic field into an alternating current. A transmitting antenna converts AC to an EM field.

Antenna detuning
Relative to RFID implementations, the reduction in the amount of energy that is available to power a tag, or the reduction of the size of the signal reaching the reader, due to the environment.

Anti-collision
See collision avoidance.

Anti-theft system
HF passive tags used on clothing and other merchandise, which cause an alarm when passed through a portal at the retail store exit.

ASN

Advance Shipping Notice, used by Wal-Mart and others to confirm shipment in advance from a supplier.

Attenuation

The loss of energy as a signal propagates outward.

Auto-ID Center

Formed at MIT in 1999 to create standards and methods for RFID, the center closed operations and passed its work to EPCglobal.

Auto-ID tags

A general term for RFID tags used in the supply chain.

Backscatter reflection

Far-field electromagnetic waves that bounce off of and propagate away from an object. With passive tag RFID, the tags modulate the backscatter reflection to create a unique response.

Bar code

An automatic identification technology that encodes information into an array of adjacent varying width parallel rectangular bars and spaces.

Bioterrorism Act of 2002

Regulations administered by the US Food and Drug Administration to safeguard the food supply. Requires registration of all importers of food for human and animal consumption. A Dec. 2004 ruling on the establishment and maintenance of records calls for track and track capability to which RFID applies.

Case

Term for an exterior shipping container within a palletized unit load or an individual shipping container.

Certified label

Labels that are approved by a printer supplier for use with their equipment.

Circular polarization

Antenna design where the energy is broadcast in a number of angles to its plane, creating a more circular pattern.

Collision avoidance algorithm

RFID reader firmware that intercepts multiple simultaneous tag signals, sorts responses, and initiates a communications protocol to sequentially collect the information.

Commercial and Government Entity Code (CAGE)

A five-position alpha-numeric code assigned by the US federal government to suppliers of goods procured by the DoD and other government entities.

Compliance

Term to describe planned activities that meet a mandate or directive.

Conductivity

The ability of a substance to carry an electrical current. Conductive materials such as liquids and metals tend to absorb radio waves (electromagnetic energy), and attenuate them.

CPG
Consumer Packaged Goods.

CRC
Cyclic Redundancy Check. The checksum calculation field in a binary coded EPC communication.

Curtain
RFID reading area on a packaging line, where reader antennas are arranged and focused to pick up tags passing between them.

Customs-Trade Partnership Against Terrorism (C-TPAT)
An outgrowth of the Patriot Act of 2001, C-TPAT is administered by the US Bureau of Customs and Border Protection. C-TPAT is designed to protect US borders against terrorist of terrorist weapons infiltration. March 2005 incentives provide benefits for RFID tagging.

DC
Distribution Center.

Dead zone
General term for an area where an RF signal cannot be read.

Defense Logistics Agency (DLA)
A branch of the US Department of Defense that over-sees procurement and supply. By some measures, the DLA is the second largest distributor in the world.

Decibel (dB)
Unit used to express the relative difference in power or intensity, usually between two acoustic or electric signals. 0 dB is equal to a one to one ratio, and 10 times the power is 10 dB.

Dense reader mode
Generation 2 operating mode to optimize reader-tag communications in environments with many readers transmitting simultaneously. Dense reader mode is a frequency channelization scheme that restricts readers transmissions to non-overlapping channels.

Dielectric constant
The measure of a material's capacitance, or its ability to store an electric charge. If a material has a high dielectric constant, it detunes an RF antenna, making it harder to read a tag. Examples of materials with a low dielectric constant are dry paper, plastics and glass.

Dipole
Antenna made of two straight electrical conductors or poles.

DoD
Department of Defense.

Dual Dipole
An antenna that has two dipoles, usually at right angles to each other, to improve sensitivity at various orientations.

EAN

European Article Number. An eight or 13-digit code originally used by companies outside North America to uniquely identify themselves and their products worldwide.

EAN International

Based in Brussels, and now called GS1, EAN International is a member organization that jointly manages the EAN.UCC system with the UCC.

EAS

Electronic article surveillance. A theft detection system where RF tags are attached to high value clothing and other items in a retail store.

ECCnet

A service of the Electronic Commerce Council of Canada, providing a secure, online, single source of standardized item data continuously synchronized with trading partners.

EEPROM

Electrically Erasable Programmable Read-Only Memory. A non-volatile storage method used in some passive tags.

Electromagnetic field

Produced when charged objects such as electrons in a wire, accelerate and decelerate. All EM fields display properties of wavelength and frequency.

Electromagnetic interference (EMI)

Energy byproducts of motors, sunspots, etc., within the RF spectrum that disrupt communications.

Electrostatic discharge (ESD)

High voltage discharge caused by a build-up of electrical potential within an object that is isolated from ground. ESD can damage integrated circuits, including RF chips. Proper handling procedures and precautions are used to limit damage from ESD.

Encoder

A reader and antenna built into a smart label printer to write information to tags.

EN 302 208

A new UHF transmission standard currently undergoing ratification in the European Union. EN 302 208 allows license-free use of more bandwidth and higher powered transmissions than prior standards, similar to US standards.

EPC

Electronic Product Code. A labeling code that identifies the manufacturer, product category and individual item. Created by the Auto-ID Center/EPCglobal, EPC is backed by the United Code Council and EAN International, the two main bodies that oversee bar code standards.

EPCglobal

An independent, not-for-profit, global standards organization entrusted by industry to drive adoption and implementation of the EPCglobal Network™ and EPC technology.

ERP

Enterprise Resource Planning. Usually refers to a software application.

Far field

The distance at about one wavelength or greater where the electromagnetic field separates and radio waves propagate away from the magnetic (near) field.

Field strength

A measure of radio signal reception.

Firmware

Software programmed into a non-volatile memory chip.

Fixed RFID reader

A reader installed at a location, with an external power supply, antenna and network connection.

Frequency

The number of repetitions of a complete waveform in a specific period of time. 1 kHz equals 1,000 complete waveforms in one second. 1 MHz equals 1 million waveforms per second.

Gain

A comparitive measure of how much electromagnetic energy an antenna will collect or emit when the antenna is focused in one direction, as opposed to emitting energy of equal intensity in all directions. Most UHF RFID systems use a patch antenna with a gain of six dBi, meaning the radiated power is multiplied by a factor of four in one direction.

Gen 2

The EPCglobal UHF Generation 2 standard is currently in review for adoption by ISO. It defines the next generation EPC encoding and air interface standards for tag to reader and reader to host communications.

GIAI

Global Individual Asset Identifier. Used by a company to label fixed inventory.

GLN

Global Location Number. Used within the EAN.UCC-13 data structure to identify physical, functional and legal entities.

Global data synchronization

A multi-industry objective and process whereby item data is represented in standard formats, and these formats are universally adopted to facilitate electronic information exchange between manufacturers, distributors and retailers.

Good label
A label is considered good when the RFID data is written to the tag correctly, the correct image is printed and content data is verified against the source.

GPIO
General Purpose Input Output interface for Printronix printers allowing control of extended devices.

GRAI
Global Returnable Asset Identifier. Used typically to track returnable containers.

GS1
Formerly called EAN International, GS1 represents 101 member organizations worldwide and administers the EAN.UCC system.

GTAG
Global Tag. An EAN.UCC initiative supported by Philips Semiconductor and others for asset tracking and logistics.

GTIN
Global Trade Item Number. An umbrella term used to describe the entire family of EAN.UCC data structures for trade items.

Half-duplex
Mode of radio communication where the same band is used both ways and transmission occurs in one direction at a time.

Handheld RFID reader
Battery powered reader used to verify tags and locate tagged items.

Hertz
A measure of electrical frequency oscillation.

High frequency tag
Tags operating in the 13.56 MHz band.

IC
Integrated Circuit.

Inductive coupling
Method of creating a current in a conductor without touching it directly to a power source. A tag responds to a reader by inductively coupling with the reader carrier signal.

Inlay
The combined chip and antenna mounted on a substrate and attached to label stock to create a smart label.

Interference
Any environmental condition that creates electrical noise at the same frequency as the communications signal.

Interrogator
Another term for a reader.

ISM bands
Industrial, Scientific and Medical government-regulated radio frequency bands.

Item
General term for an individual product or service.

Kill Command
A password protected instruction that renders a tag inoperable.

Linear polarization
Antenna design where energy radiates mostly in a straight line pattern at 90 degrees to its plane.

Line-of-sight
Technology that requires an item to be "seen" to be automatically identified by a machine. Bar codes and optical character recognition are two line-of-sight technologies.

Low-frequency tag
Tags that operate in the 125 KHz band.

Microchip
A microelectronic semiconductor device comprising many interconnected transistors and other components. Also called an IC.

Microwave tag
Tags that operate in the 5.8 GHz band.

MIL-STD-129P
Specification for marking of packages shipped to the US military. MIL-STD-129P w/Change 3 details RFID labeling requirements.

Nano-technology
Integrated circuit technology where circuit size can be expressed in nanometers, a unit of which is only 3-5 atoms wide.

Near field
Portion of the electromagnetic field, within a distance of one wavelength, where the magnetic field follows a flux line path.

Network print management
A system of printer/encoders communicating through a host computer interface, allowing centralized control and visibility into the enterprise printing function.

ONS
Object Name Service. Similar to the Domain Name System associated with the World Wide Web, ONS is an Auto-ID Center designed system for looking up unique EPC through an Internet linked computer.

ODV™
Printronix Online Data Validation.

Opaque
Relative to RFID, an opaque object does not allow a reader signal to pass through it.

Order cycle time
The speed at which a retail location can replenish stock. Used as a measure of supply chain efficiency.

Out-of-stock
Retail term for product offered for sale but not in inventory.

Overstrike
Printer capability to recognize a bad label, back up and mark it as such by printing a grid on top of it.

Pallet
General term for a skid full of cases that can be handled by a fork lift.

Palletized unit load
Military term for a skid full of cases that are secured, strapped or fastened on a pallet and labeled according to MIL-STD-129P. A palletized unit load is distinct from a shipping container.

Parametric test
Test of a tag inlay after assembly to detect quiet tags.

Passive RFID tag
Passive tags do not have an on-board powered transmitter. They are activated by the electromagnetic waves of a reader, with a read range of up to 25 feet.

PGL™
Printronix Graphics Language. Native language of Printronix printers, for programming graphics and control commands.

Pilot
The stage of an implementation where technology, process and methods are evaluated before committing to their use in a production environment.

PML
Physical Markup Language. An Auto-ID Center designed method of describing products in a way computers can understand. Based on XML.

Polarization
The orientation of flux lines in an EM field.

Portal
A defined RFID reading area, where readers are mounted specifically to read tags going through, such as a dock door or over a packaging line.

POS
Point Of Sale.

Power over Ethernet (PoE)
Readers and other network devices equipped with PoE use the Ethernet CAT5 cabling to draw electrical power, saving a separate power connection. This allows greater flexibility in locating network devices and reduced installation costs. A PoE router upstream in the network is required.

Print and apply
Automated approach to printing, encoding and applying smart labels in a packaging process.

PrintNet® Enterprise

Printronix networked print management system.

Print quality

The measure of compliance of a bar code symbol to ANSI and/or traditional requirements of dimensional tolerance, edge roughness, spots, voids, reflectance, quiet zone and encodation.

Quiet label

A label that cannot be read from a normal distance.

Radiolucent

The degree to which a substance absorbs or transmits radio waves that attempt to pass through it.

Radio waves

Waves at the lower end of the electromagnetic spectrum.

Read after print

A step in producing smart labels, where the printer interrogates the RFID tag and reads the bar code to verify a good label.

Reader

Also called an interrogator. The reader communicates with the RFID tag and passes the information in digital form to the computer system.

Read only memory (ROM)

A form of storing information on a chip that cannot be overwritten.

Read only tag

Encoded during the tag manufacturing process so information cannot be changed. EPC Class 0.

Read range

The distance from which a reader can communicate with a tag. Range is influenced by the power of the reader, frequency used for communication and the design of the antenna.

Read rate

The accuracy of a bar code or RFID system, expressed as a percentage of good reads versus bad. May also be a benchmark or theoretical rate.

Read redundancy

The number of times a tag can be read while in the read window.

Read write tag

A programmable tag, EEPROM or battery-backed memory. EPC Classes 0+, 1-4, including UHF Gen 2.

Receiver

Another term for a reader, as in radio receiver.

RIED

Real-time In-memory Event Database.

Replenishment

Supply chain term for ordering and receiving stock.

RF

Radio Frequency.

RFID

Radio Frequency Identification. A method of tracking using radio waves that trigger a response from a device attached to an item.

RFID Printer/encoder

An RFID printer that encodes the smart label and immediately checks to verify if the tag is readable. Same as smart label printer.

ROI

Return on Investment, usually measured as capital equipment and labor costs offset by tangible and intangible returns over a given period of time.

Sarbanes-Oxley Act

Legislation requiring US public companies to follow financial reporting rules as administered by the Securities and Exchange Commission. Section 404 of SOX relates to the integrity of financial records, which can include RFID data.

Savant

Auto-ID Center term for distributed network software that manages and moves data related to Electronic Product Codes. The term has been retired in favor of ALE.

Semi-passive tag

Passive tag that uses battery-power to boost its transmission range.

Serialization

The unique numbering of objects for identification purposes.

Shielding

Materials that have a significant electrical conductivity. Shielding may be used to directionally orient a signal, such as the wire braid jacket on an antenna cable. Objects that are electrically conductive at high frequencies, such as aluminum cans, may cause shielding that is detrimental to the performance of RFID.

Shipping container

An exterior container which meets carrier regulations and is of sufficient strength to be shipped safely without further packing.

Shrinkage

Retail term for product loss after receipt from a supplier, often due to employee theft.

Singulation

The act of isolating one tag within a tag population. Tag sorting algorithms are used by readers to singulate a tag before initiating a query to obtain the EPC number. When initially programming a tag, singulation occurs by physically isolating it from others, usually by a printer/encoder using near-field coupling.

SKU

Stock Keeping Unit.

Slap & Ship
Supply Center language for a post-process operation usually in a distribution center by which goods requiring RFID labeling are delivered and tagged just prior to shipping.

Sleep mode
An ID tag command to suppress an RF response in a tag. Used to minimize unintended reads.

Smart label
A label that contains an RFID tag. It is considered "smart" because it can store information, such as a unique serial number, and communicate with a reader.

Smart label printer
RFID printer that produces smart labels.

SSCC
Serial Shipping Container Code. Used to identify cases, pallets and other containers.

Strap
Section of an RFID chip surrounding the chip, to which the antenna is attached.

Supply chain
Industry term for a group of companies working together to manufacture, inventory and supply materials to fulfill a production schedule or finished products to meet market demand.

Tag
The generic term for a radio frequency identification device.

Thermal printer
A device that uses a heat transfer process to print characters or graphics, for continuous roll printing of tape or labels.

TMS
Task Management System.

Tote
A re-usable container that can be hand-carried.

Trade item
General term for a product or products packaged in lot sizes and made available for transport and sale.

Transmitter
Portion of a reader that includes the antenna and is used to broadcast a radio signal.

Transponder
Refers to the part of an active (battery-powered) tag that includes the antenna and is used to emit a response to a signal.

UCC
Uniform Code Council, now replaced by GS1.

UCCnet

An Internet-based product registry service for standards-based electronic commerce. A subsidiary of GS1 US, the service enables synchronization of item and location information among trading partners.

UHF

Ultra-High Frequency. The term generally given to waves in the 300 MHz to 3 GHz range. UHF offers high bandwidth and good range, but UHF waves don't penetrate materials well and require more power to be transmitted over a given range than lower frequency waves.

UID

Universal Identification, code used by USA Department of Defense to mark and track assets that require serialization.

UPC

Universal product code. The bar code standard used in North America over the last fifteen years. It is administered by the GS1 US.

Wavelength

The inverse of frequency, a wavelength is the measure of peak to peak of a radio wave.

WMS

Warehouse Management System.

XML

Extensible Markup Language, for defining, validating and sharing documents containing structured information. Unlike HTML, XML tags can be designed for specific purposes.

CHAPTER 1

Introduction

REVOLUTIONARY TECHNOLOGY, OR COST BURDEN? 3

Reducing Inventory Guesswork 4

Why RFID Now? 5

Using the Supply Chain for Competitive Advantage 7

Much More than a Supply Chain Mandate 8

BUILDING A BUSINESS CASE FOR RFID 11

Pushing RFID Down the Supply Chain 16

Hype and Misconception 18

PROTECTING INVESTMENTS DURING IMPLEMENTATION 20

HOW THIS BOOK IS ORGANIZED 22

Lessons Learned 23

IN THIS CHAPTER:
Business issues
that are driving the
adoption of RFID

When Wal-Mart announced in late 2003 an initiative for RFID use within its supply chain, suddenly the rules changed. Wal-Mart's top 100 suppliers were being asked to invest and re-engineer in order to retain their status as suppliers. Other US and European retail companies started similar RFID initiatives, including Albertsons, Carrefour, Metro, Target, and Tesco. The US Defense Department issued a similar directive. Before anyone could get used to the idea, thousands of companies fell under some sort of RFID compliance initiative.

It is fair to ask what got us here. Why did all of these companies get served with a mandate? What's wrong with their current bar coding system? What is it about RFID that is so compelling that Wal-Mart, the DoD and others will risk potential disruption within their supply chain? After all, this isn't just slapping a tag on a case or pallet. In order to implement RFID, you must have systems in place to serialize tags on cases, pallets and eventually every item you produce and ship. Then you need systems to read, track and leverage value out of all of that data. Finally, you need systems to synchronize data-driven processes within the supply chain. That's a significant change.

Can RFID help companies solve real business problems? Can it truly improve inventory tracking, product shrinkage and out of stock at the retail store level? The basic premise of RFID is that by attaching a radio frequency tag to an item, a computer can track that item without human intervention. By tracking an item remotely through key events in its "life," you can automate its flow through the supply chain. Business rules written into ERP and WMS software can guide its flow, from raw materials to retail shelf, from factory to fox hole. Presumably, RFID will help supply chains tune themselves to respond more quickly to consumer demand.

If RFID can do this, then compliance is suddenly not the issue. Business survival is the issue – staying ahead of the curve, the coming revolution, where only the RFID-enabled supply chain companies succeed.

REVOLUTIONARY TECHNOLOGY, OR COST BURDEN?

The question is not whether RFID is disruptive, rule-changing technology. Compliance mandates took care of that. The real question is, can RFID deliver enough economic benefits to transform the way supply chains do business and recoup the upfront costs? Will this be true for every supplier? How quick is the payback?

Like the personal computer, the fax machine, the Internet, and bar codes in their inception, RFID has the potential to transform commerce. RFID promises to take people out of the identification and data collection loop. A recent study by IBM estimates that RFID could reduce labor involved in the receipt of goods by 60 to 90 percent. The results of recent implementations are sharpening the picture. Proctor & Gamble reduced the number of forklift drivers at a manufacturing plant in Spain after introducing an RFID system. Ford Motor claims a 10 percent labor savings at its Sterling Heights, Michigan manufacturing plant.

Reducing Inventory Guesswork

Labor savings is not where the bigger benefits of RFID lie. Rather, bigger benefits come from RFID's potential for solving a data-availability problem. RFID improves product availability at the retail level without adding inventory by helping companies better track, automate the flow of and understand the condition of goods in the supply chain.

→ SEE PAGE 138
on Wal-Mart.

A recent study by the Grocery Manufacturers of America (GMA) found that consumers cannot find the product they are looking for 8 percent of the time. Advertised product promotions cause nearly twice the number of out-of-stock experiences. When shoppers find a product is out of stock, they either postpone their purchase or shop elsewhere. In aggregate, this puts $6 billion in sales in limbo, according to the study. Wal-Mart estimates that RFID will allow it to recapture one percent of revenue by improving out-of-stock, which translates to about $2.5 billion. A.T. Kearney estimated the average out-of-stock improvement using RFID to be 3-5 percent.

Excess inventory is the flip side of the coin. Some experts estimate that 30 percent of inventory in the supply chain is buffer stock – it exists because demand and supply information is not precise and real-time. GMA estimates that unsalables for warehouse-delivered consumer package goods cost the industry $2.6 billion per year in the US. A recent consumer packaged goods study estimates that $5 billion of obsolete goods are written off each year.

Why RFID Now?

Can we predict the timing of RFID adoption? We have come a long way since the can opener, but we may not be any better than our ancestors at rewarding innovation.

The story of the can opener might sober any irrational exuberance about RFID. First came the tin can, patented by an Englishman, Peter Durand, in 1810, and adopted by Napoleon as a way for pre-serving food for his army ('An army marches on its stomach.'). A manufacturing process was developed in 1846 to mass produce cans at a rate of 60 an hour. But it wasn't until 1866, over fifty years after the can was invented, that an American came up with the idea of a can opener. He still had to give it away for free with cans before it was finally adopted. For fifty years people didn't seem bothered by having to use a hammer and chisel to enjoy French food.

RFID is like a can opener. It derives its value within a well-defined system, it is clearly an improvement over the previous method, but people still don't want to pay for it.

Four factors determine the timing of technology adoption, and help answer the question of why a company should consider RFID now:

→ SEE PAGE 25
for radio frequency
standards and Page 83
for EPC standards.

Standards – Global, pan-industry standards-making organizations are taking an active role. There is strong consensus, although there are many issues yet to be ironed out, and it is probable that standards variations will exist for the foreseeable future.

Costs – Costs for tags and readers are declining and are expected to decline even further in the coming years. A year ago, a 4x6 inch smart label cost about $0.50. The price has dropped nearly fifty percent since then.

Multiple suppliers and vendors – A number of well-funded product companies, with supply chain expertise, have announced products. There is competition in the market for both products and services for system integration and achieving compliance. It is expected that the RFID hardware market (tags, readers and printers) will grow to $5 billion by 2009 at a 41 percent compound annual growth rate.

Leadership – Wal-Mart, DoD and other leading companies and opinion leaders are committed to RFID. These leaders are asking their supply chains to formally commit to RFID through their compliance initiatives.

→ SEE PAGE 137

on industry initiatives and Page 161 on the DoD.

Using the Supply Chain for Competitive Advantage

Companies such as Dell and Nokia see their supply chains as a basis for competitive advantage in the market. These companies and others have reduced costs and increased speed and agility in their supply chains, pulling customers away from their competitors. Stanford University researchers found that leaders and transformers in the supply chain performance category also led their respective industries in terms of growth rates of market capitalization over the last ten years (Fig. 1.1). To be effective at leveraging the supply chain, according to the study, requires a company to harness information,

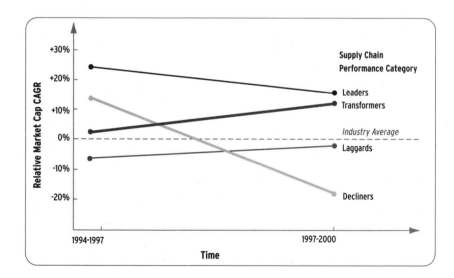

FIGURE 1.1

Supply chain leaders and transformers lead their industries. (Based on a joint study by Stanford University, INSEAD and Accenture, 2003).

and use a six sigma approach to measurement, identification, analysis and action.

97 percent of those responding to a recent Fortune 500 survey plan to deploy better information technologies to address supply chain problems. The top two technologies mentioned in the survey were RFID and supply chain management software. These technologies are expected to help companies be more efficient in forecasting demand and in using transportation. This is in light of a looming transport industry capacity crisis, where trade outstrips both carrier capacity and the infrastructure at shipping ports, especially on the West Coast of the US. Technology will also help address issues related to supply chain integrity and security, which is increasingly being mandated by legislation.

Much More than a Supply Chain Mandate

RFID adoption began fairly simply, with mandates to top tier retail and DoD suppliers. Going forward, adoption has the look of a perfect storm. See Figure 1.2. The economic forces extend through the retail and military supply chains. The technology forces extend through standards making organizations, radio spectrum regulators, chip designers and software providers. Societal forces are also there. Their roots are in the September 11, 2001 terrorist attacks on the World Trade Center in New York, and the outbreak of "Mad Cow" disease leading to the banning of potential prion-tainted animal feeds in the

US in 1995. The forces grew with public distrust of corporate financial reporting which gave birth to the Sarbanes-Oxley Act.

Today, RFID is used to tag cattle in Canada, Australia and other parts of the world. RFID tags are used to seal shipping containers under the guidelines of the US Bureau of Customs and Border Protection. And with SOX Section 404 calling for increased internal controls on

→ SEE PAGE 152
on protecting the global food supply.

→ SEE PAGE 154
on the challenges of ocean shipping.

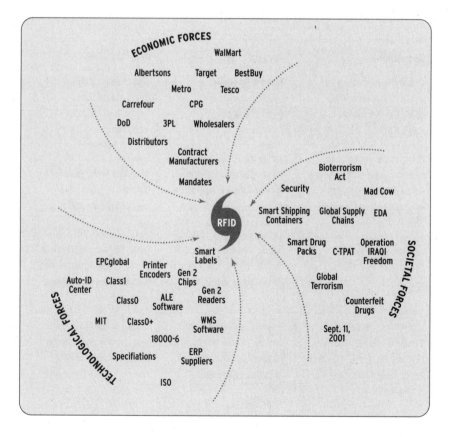

FIGURE 1.2

RFID adoption momentum looks like the perfect storm.

company assets, RFID data becomes part of a company's fiduciary process. These societal forces are perhaps more far-reaching than the others. They stem from concerns about the safety of the drug and food supply (Bioterrorism Act), the security of borders (C-TPAT) and the integrity of our financial markets. Table 1.1 lists those regulations that are driving the adoption of RFID beyond the realm of retail mandates.

→ SEE PAGE 253

on regulatory and operational requirements.

TABLE 1.1

Regulations driving RFID adoption in the global supply chain and beyond.

Regulation	Purpose	RFID Potential
Customs-Trade Partnership Against Terrorism (C-TPAT), new guidelines published March 2005.	Securing US borders against terrorists and terrorist weapons.	Smart tags on shipping containers can earn "green lane" clearance at US ports.
Public Health Security and Bioterrorism Preparedness and Response Act, 2002, Section 306 (published in December 2004).	Secure the human and animal food supply. Requires offshore supplier registration, records retention and rapid response to FDA calls for action.	Smart tags help quickly trace back to the source of contamination and trace forward to locate tainted supply.
Sarbanes-Oxley Act, Section 404.	Internal controls accuracy, records retention and integrity of financial reporting.	RFID data becomes transaction record for raw materials, WIP, finished goods inventory and goods in transit.
FDA policy, "Combating Counterfeit Drugs," published February 2004, and "Compliance Policy Guide," published November, 2004.	Protect the US Drug Supply from counterfeit drugs through chain-of-custody records and electronic pedigrees.	Tagging of retail prescription drugs as they move through the supply chain, by 2007.

BUILDING A BUSINESS CASE FOR RFID

For a company that supplies to a major retailer or the DoD, the case for implementing RFID may appear straightforward, but that's only on the surface. The business case has many layers. It is a case for the boardroom as well as the shipping dock. The value proposition is different for every company. Not all companies will benefit from RFID equally. Some will see significant returns. Others will see very few, if any at all in the near term. Issues for suppliers to consider include:

⊙ **SEE PAGE 182**

on feasibility studies.

Risk of non-compliance – Figure 1.3 illustrates some of the issues. Near term, you stand a chance of losing business if your customer insists that you comply with an RFID mandate and you don't. If the customer represents a large percentage of your business, that is sufficient justification for some level of compliance activity. If the customer is a relatively small part of your business, the risk of non-compliance may be acceptable when compared to the cost of implementing RFID. The yearly profit of that piece of business may not be anywhere near the up front costs of RFID. You could elect to forego that business in the near term, implement RFID if necessary at some point in the future, and risk later getting back in good graces with your customer. The issue changes, however, if other companies in the retail industry and elsewhere announce RFID mandates, and you are a supplier to them as well.

⊙ **SEE FIG. 1.3**

Compliance by outsourcing – Outsourcing your compliance to an RFID-enabled wholesaler, distributor or third party logistics (3PL) provider might be the lowest risk approach. Simply ask companies like FedEx, UPS, and DHL to handle it for you. You pay for a new class of service, like choosing 1-day instead of 2-day delivery. 3PLs can insure that your product meets your end-customer RFID tagging requirements. What's more, by using their electronic reporting mechanisms, you can obtain the same instantaneous customer information that you would from your own. If you rely on 3PLs for material procurements, and you source from other countries, you can use them to help implement RFID in your own supply chain.

FIGURE 1.3

Assessing the risk of non-compliance.

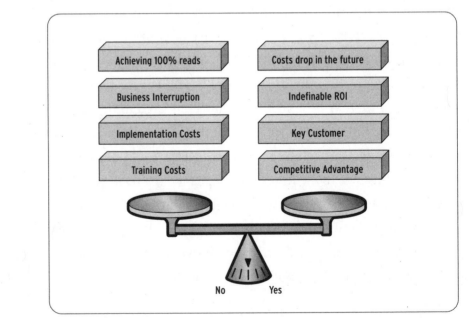

Global supply chains are affected by a myriad of regulations, and an RFID-savvy 3PL provider can play an essential operational role in your business by providing visibility to inventory, and instantaneous reconciliation of shipments to receipts.

High-volume, low-margin product businesses – Examples of such products are soaps and cleaners, frozen foods, perishables, everyday wear, etc. If a high percentage of your business is affected, and you produce or re-package low-margin products in high volume, the cost of compliance is considerable, and could effect your overall profitability of doing business. An RFID tag will be a significant cost adder to each case of product shipped (Fig. 1.4). Today's implementation cost-levels, according to one analyst, make RFID attractive for products at or above a $15 retail cost. Products below that cost need a more rigorous business case justification.

→ SEE PAGE 318 food iudustry case studies.

→ SEE FIG. 1.4

With low margin products, a case could be made for an aggressive and thorough implementation to achieve maximum direct labor savings and efficiencies from RFID as soon as practical. On the other hand, if a relatively small portion of your business is affected, a business case could be made to reducing exposure to RFID implementation costs by outsourcing the RFID compliance work, or diverting production to an area of your distribution center where it can be re-worked by hand. By avoiding the high costs of re-engineering all your processes for RFID in the near term, you may gain in the long run should implementation costs drop drastically as RFID moves toward mass adoption.

Low to medium volume high-margin product businesses – Examples include bicycles, cosmetics, video games, electronics, etc. For computer peripherals, televisions, and other big-ticket products, where the case equals the item-level, RFID and EPC implementation could bring near immediate benefits. Item level serialization is already being done, and RFID tracking could improve out-of-stocks and item shrinkage, which are significant cost issues. See Figure 1.5. Internal implementation, rather than outsourcing to a contract

→ **SEE PAGE 338**

electronics industry case study.

FIGURE 1.4

Tag cost as a component of product cost.

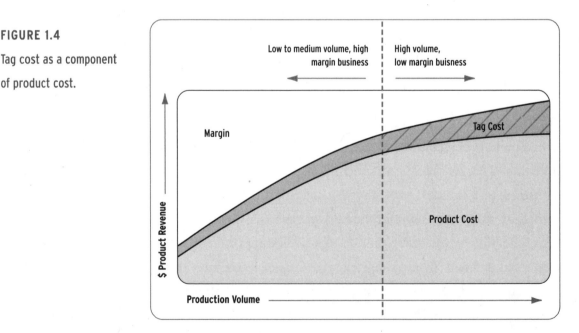

RFID compliance center, may be the best choice.

Products with a long lifecycle, especially those that require serial number tracking, warranty service and spare parts inventories, are good candidates for a staged implementation that goes beyond compliance toward extensive business-process integration. A number of trial options are available, including manual tagging and diverting product to semi-automated lines.

Complex supply chain – Companies that supply products direct to store, such as perishables (flowers, bakery goods), contract bottlers or jobbers for snack items, should look to their customer

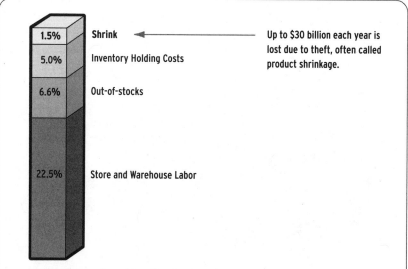

Retailer Average Annual Benefits (% of each $1 of sales)

FIGURE 1.5

Big-ticket items are more affected by out-of-stocks and shrinkage.

and supply chain partners for assistance. It may be possible to install RFID tags on re-usable containers such as totes and pallets. Direct to store supply chains may require significant cost sharing among partners.

Pushing RFID Down the Supply Chain

A business case for RFID may require a hard look at your own supplier practices. Depending on your business, the direct benefits "within the four walls" may not be as significant as those gained by having your suppliers and trading partners implement RFID (Fig. 1.6). If you are a volume-driven manufacturer who dominates your industry, you stand to gain considerably by pushing RFID down to suppliers.

FIGURE 1.6

RFID benefits gained through supply chain and trading partners.

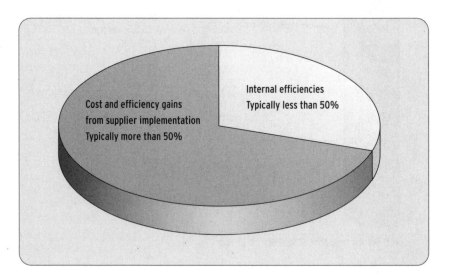

It is fair to say that benefits of RFID will not accrue at the same rate, and in the same fashion, in each link of the supply chain. The organization at the top of a supply chain – the retail store, the troops at the front in the case of a military supply chain – will gain most quickly by elimination of out-of-stocks. Other suppliers in the chain, however, will also see gains through increased revenues by having their products available on the retailer's shelves. Companies that produce big-ticket items, manage high value assets, or control the point of sale, and that implement RFID systemically, will benefit quickly as well. Everyone else will have to wait for the trickle down of transparent, instantaneous data in the supply chain, and the labor savings as RFID pilots mature. It might make sense to consider RFID in two time frames:

Near term, compliance driven – Integrate RFID into your case and pallet supply systems with as little downtime and business interruption as possible. One obvious path is by converting your existing UPC bar code labeling stream to EPC smart labels.

Long term, ROI driven – Use RFID to change the rules in your own supply chain and business. Maximum value will come when you've taken the opportunity to re-think your processes. Look for artifacts of legacy material handling systems based on old technology and restructure them. Eliminate procurement and manufacturing planning based on inventory buffering and cueing. Comprehensive changes such as these should only be undertaken when you have internalized all the challenges and potential obstacles for successful RFID technology implementation, and can engineer appropriate solutions.

Hype and Misconception

→ SEE PAGE 179

on getting started.

There has been enough practical knowledge built up over the last couple of years to help any company take a disciplined approach to RFID. You start by assembling an RFID team, conduct a feasibility study, start a pilot and build a business case for adoption. Most teams can quickly dispense with the hype and misconception surrounding RFID, and move forward. But as more people get involved at a company, a good education program is needed to orient people for positive change, and dispel these myths:

→ SEE PAGE 106

on addressing consumer concerns.

Spy chips – Privacy groups are actively campaigning against the use of RFID tags on consumer items, claiming that they are invasive and could potentially erode consumer privacy. To combat misinformation about RFID, industry groups have adopted visible labeling and usage guidelines, and have started public educational programs. Passive UHF tags, by and large, have a read range of less than 30 feet, and their use within the supply chain does not constitute an invasion of privacy or an attempt by anyone to spy on private individuals.

Data banks – Generation 1 chips have up to 128 bits of memory. Gen 2 will typically have 512 bits, or 32 words. That's less than your average English language sentence. RFID chips were never designed to store all the information about a product. They store serial numbers that have no value by themselves. If someone were to read and decipher the contents of a chip, they would need access to supply chain databases, which are typically secured, in order to obtain descriptive data about the item, originator, destination, cost, etc. A thief looking for valuable data is better off standing on a street corner and writing down vehicle license plate numbers.

→ SEE PAGE 96
on data storage.

Read and write on the fly – Tags pretty much read on the fly but they certainly don't write that way. The tag programming range is 30 percent less than its read range typically. It needs to be programmed first before it will respond to subsequent "on the fly" programming commands. Most passive UHF tag implementations use a printer/encoder to initially program the tag, and thereafter use the tag in read-only mode. Companies are more concerned about getting the tag to read reliably at each point in the supply chain, knowing that they can use external databases to keep track of its status.

→ SEE PAGE 121
on encoding smart labels.

Bar code replacer – RFID has advantages over bar code to speed the movement of goods and create supply chain efficiencies. Where it meets these goals, it will replace bar code and other identification and tracking technologies. RFID, however, will never replace visible labels and package markings that aid human workers in identifying contents and doing work. Bar coding has its limitations, but as long as it continues to be used in consumer packaging, and the bar coding infrastructure in place provides value, it will continue for years to come.

⊙ SEE PAGE 45

on Gen 2 specification.

Gen 2 compatibility – The standards making process in the RFID industry is progressing toward a goal of "harmonization." Vendors and customer involved in the standards making process know that it is in the best interest of end-users to have device interoperability, multiple sources of supply, and compatibility with earlier generation product. This reduces the risk of investment and moves the market forward. Like many community-driven decisions, however, the Gen 2 specifications sacrificed some functionality for the sake of consensus. The specifications leave room for vendors to provide enhancements and special features that differentiate their products and possibly provide their customers competitive advantage. Careful screening and selection of vendors and products, and interoperability testing, will continue to be important in the industry.

PROTECTING INVESTMENTS DURING IMPLEMENTATION

Any business case for early adoption will be helped by the RFID team making smart choices about approaches, equipment and vendor partners. Equipment and system integration will be the two largest investments. If they are kept to one-time costs, returns tend to accrue sooner. Here are some guidelines to consider:

Upgradeable devices – RFID printers and readers that are firmware upgradeable will retain their use through the lab, pilot and production phases of an implementation. Look for devices that are engineered specifically for rugged use in production environments, device suppliers with proven industry experience and a commitment to helping you protect your investment. This is especially important as compliance requirements shift to the UHF Gen 2 standard starting in 2006.

Software migration tools – Device vendors should provide migration tools that support the conversion of data streams set up for UPC bar coding over to EPC. This includes capabilities to manage the assignment of EPC serial numbers across multiple operations producing smart labels.

Certification of labels and tags – As the single biggest cost item, you will need flexibility in sourcing labels. Look for vendors who offer label compliance and certification services, that work closely with printer/encoder suppliers to guarantee the quality of the encoding printer, and professional services teams who will work with you directly to help you achieve compliance.

⊙ SEE PAGE 119
on label certification.

System integration and automation – Vendor teams who have RFID and supply chain experience working together on similar projects offer the best assurance of success as you move toward volume production. Many of these system integration teams tend to be from smaller firms that specialize in retail packaging applications. Their experience and ability to work with both legacy systems and open-source languages and protocols will help ensure against having to scrap a system and start over.

Backup and recovery capability – Smart labels with bar coding offer the best way to identify and recover from RFID problems. Look for ways to streamline RFID into your current bar coding process, and partners that will help you retain the integrity of both systems for the foreseeable future.

HOW THIS BOOK IS ORGANIZED

This book covers a number of related subjects important to understanding and using RFID and smart labels in particular. It is not necessary to read each chapter in sequence. Chapters are organized in subjects that cover various layers of an RFID system, from the device layer (Chapter 2), to the application integration layer (Chapter 10). Figure 1.7 shows the layers and related chapters. At the back of this book is a supplementary section of products from Printronix.

Lessons Learned

Throughout this book, we'll flag what RFID practitioners have learned from actual implementations. Early participants in the DoD and retail RFID trials have had a year of discovery and implementation, and are very willing to share their lessons.

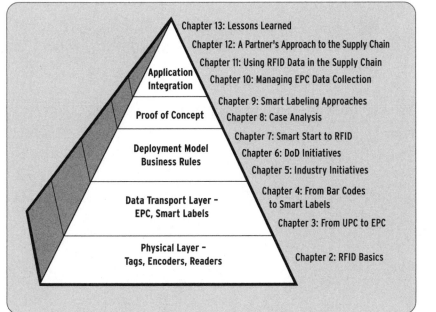

FIGURE 1.7

Organization of the book.

Sources and Further Reference

Meeting the Retail RFID Mandate, white paper, A.T. Kearney, November 2003.

Tag, Trace, and Transform, Launching Your RFID Program, white paper, Deloitte Touch Tohmatsu, 2005.

"RFID Hype Collides with Reality," by Dennis Gaughan, *Supply Chain Management Review,* March, 2005.

"Man vs. Machine," by Samuel Greengard, *RFID Journal,* October 2004.

Corporate Strategy for the New Millennium, white paper, IBM, available: www.ibm.com.

RFID-Read My Chips!, Research Note, Piper Jaffray Equity Research, April, 2004.

"Closing the Loop – Six Sigma Supply Chain Management," presentation by Hau L. Lee, Stanford University, *Xtended World Conference* 2005.

Fortune 500 Companies' views on the supply chain and logistics landscape 2005/2006 Report, available: www.eyefortransport.com.

CHAPTER 2

RFID Basics

RADIO WAVES 26

European Frequency Regulatory Standard 30

THE RFID TAG 32

Tag Types 37

Data Storage Capability 40

Evolution of Tag Classes 41

Generation 1 Tag Specification 43

Generation 2 Tag Specification 45

Tag Selection and Readability 49

PRINTER/ENCODERS 52

READERS 57

First Generation Basic Operation 59

Gen 2 Basic Operation 61

READER ANTENNAS 63

INTEROPERABILITY TESTING 69

THE GEN 2 STORY 70

IN THIS CHAPTER:

Parts of an RFID System

A typical RFID system consists of four main components: tags, an encoder, readers, and a host computer. See Figure 2.1. The RFID tag is made up of a microchip and a flexible antenna encased in a plastic-coated inlay. The encoder is used to write information to the tag. In coming years you might find RFID tags built into products and product packaging. For now, the most common format is a shipping label with a built-in tag, or smart label. Smart labels can be printed and placed on each case or pallet.

To get an RFID tag to "talk back" a reader broadcasts a radio wave. If it is within the range of the reader, the tag answers, identifying itself. Tags can be read from a distance without physical contact or line of sight. The distance within which a reader can communicate with a tag is called the read range. Communications between readers and tags are governed by protocols and emerging standards, such as the UHF Generation 2 (Gen 2) standard for supply chain applications.

RADIO WAVES

When electrons move, they create electromagnetic waves that can move through air. The waves can pass through some physical objects, and even a vacuum. The number of oscillations per second of an

FIGURE 2.1 Main components of an RFID system.

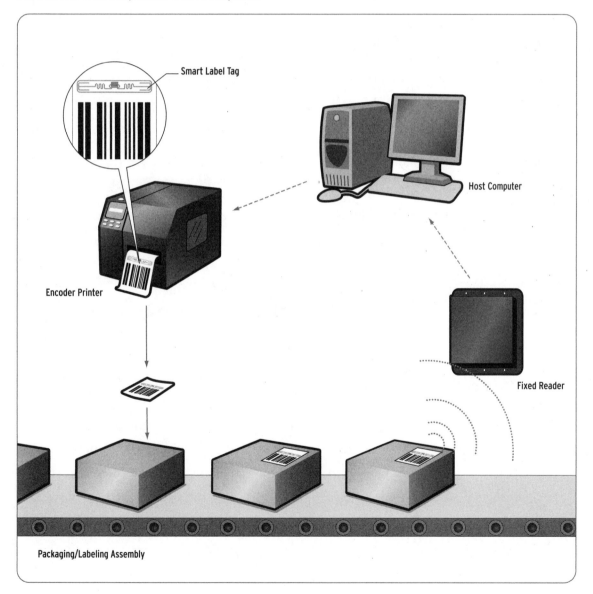

Smart Label Tag

Host Computer

Encoder Printer

Fixed Reader

Packaging/Labeling Assembly

electromagnetic wave is called its frequency, and is measured in Hertz (Hz). The distance between two consecutive wave peaks is called the wavelength.

By attaching an antenna of the appropriate size to an electrical circuit, the electromagnetic waves can be broadcast efficiently and received by a receiver some distance away. All wireless communication is based on this principle.

Radio is usually associated with long distance communications. In the case of RFID, we focus on the characteristics of radio waves over a relatively short distance. Electromagnetic waves travel through a vacuum at the speed of light. In copper wire, the speed slows to about two-thirds of this value and becomes somewhat frequency dependent. The electromagnetic spectrum is shown in Fig. 2.2. The radio, microwave, infrared and visible light portions of the spectrum can all be used for transmitting information by modulating the amplitude, frequency or phase of the waves.

The properties of radio waves are frequency dependent. At low frequencies, radio waves pass through obstacles well, but the power falls off sharply with distance from the source. At high frequencies, radio waves tend to travel in straight lines and bounce off obstacles. They diffract at corners, sharp edges, and openings.

Radio waves are subject to interference from a variety of sources, from sun spots to other electrical equipment. Radio communications has its uses in all sorts of applications, but only if interference between users can be kept at a minimum. For this reason, governments tightly license users of radio transmitters, with the exception of the industrial, scientific and medical (ISM) bands. Transmitters using these bands do not require government licensing. One band is allocated worldwide: 2.400–2.484 GHz. In addition, in the United States and Canada, bands also exist at 866–956 MHz and 5.725–5.850 GHz. These bands are used for cordless telephones, garage door openers, wireless hi-fi speakers, security gates, etc. See Table 2.1.

(→) SEE TABLE 2.1

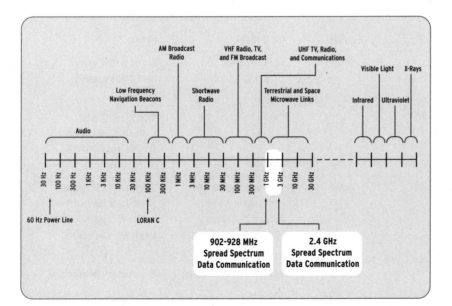

FIGURE 2.2
Portion of the electro-magnetic spectrum associated with RF communications.

Many parts of the world have specific frequencies assigned to various applications, from cell phones to security gates. This hinders the development of a single global standard for supply chain RFID use. In the United States, the FCC has long authorized the 915 MHz band for RFID use. The 900 MHz band in Europe, for example, is currently used by GSM cellular phones. Recently, Europe has adopted 866 MHz for their RFID standard. Japan is currently considering standardization at 950-956 MHz.

European Frequency Regulatory Standard

European Telecommunications Standards Institute (ETSI) has proposed additions to the current standards to improve the viability of

TABLE 2.1

For RFID, these bands have the following characteristics:

Band	Frequency	Read range	Notes
LF	100-500 kHz	Up to 20 inches (50.8 cm)	Access control, animal identification, vehicle key-locks.
HF	13.56 kHz	Up to 3 feet (1 meter)	Access control, smart cards, item level tagging, libraries, and electronic article surveillance.
UHF	866-956 MHz	FCC allows over 20 feet (6 meters) at 915 MHz. Range at 866 MHz is about 10 percent less than at 915 MHz.	Supply chain use, baggage handling and toll collection. Wal-Mart is accepting RFID tags in this spectrum.
Microwave	2.45 GHz	3 to 10 feet (1 to 3 meters)	Item tracking, toll collection.

RFID in supply chain management applications. The 46 national communications authorities in Europe voted to adopt the new regulations, which will allow RFID readers to use more power and operate in a wider UHF band. The new regulations, EN 302-208, will allow European RFID readers operating in the UHF band to perform nearly as well as UHF readers operating under FCC rules in the United States.

The existing EN 300-220 frequency allocation (see Table 2.2) provides a frequency range from 869.4 to 869.6 MHz in which RFID readers can operate. Broadcasts are restricted to a single channel, and power restrictions are such that broadcasts of greater than 5mW is limited to a 10 percent duty cycle. The new frequency allocation, EN 302-208, provides an additional frequency range, from 865 to 868 MHz, in which RFID fixed readers can operate. It supports frequency hopping, which facilitates reading multiple tags in a noisy environment. A characteristic called Adaptive Frequency Agility (AFA), requires that readers listen first before transmitting on a

TABLE 2.2

Frequency allocation regulations in various countries and regions.

	North America	Europe 300 220	Europe 302 208	Japan (pending)	Korea (new)	Australia	Argentina Brazil Peru	New Zealand
Band Size	902-928	869.4-869.6	866-868	950-956	908.5-914	918-926	902-928	864-929
Power	4W EIRP	0.5W ERP	2W EIRP	4W EIRP	2W ERP	4W EIRP	4W EIRP	0.5-4W EIRP
# of Channels	50	1	10 +5	12	20	16	50	Varied

channel to prevent clashes. A reader can transmit for 4 seconds then must either stop for 0.1 seconds to allow other devices to use the channel, or switch to another of the 10 available channels.

Standards making tends to be a complex and lengthy process, and the European RFID frequency allocation standard is no exception. Even though the new standards are well documented, each EU member now has to pass local legislation to permit the new standard. This process is underway and is expected to be completed in 2005. In the meantime, tag and reader vendors can sell equipment which conforms to the EN 302-208 or EN 300-220 regulations if they formally apply to each EU country to sell the products and the equipment is verified by an accredited testing agency. Tags encoded by a printer encoding to EN 300-220 are easily read by fixed readers confirming to either EN 302-208 or the FCC regulations.

THE RFID TAG

An RFID tag comprises a microchip mounted on a flexible PET substrate with an attached antenna. This "inlay" assembly is then "converted" or sandwiched between a label and its adhesive backing. See Figure 2.3. The actual chip may be no bigger than a grain of sand (about 0.3 mm^2). Chips used in RFID tags may become the most widely-used commercial application of nano-technology. Although the chips are tiny, the antennas are not. They need to be

(!)

The expected demand for tags is huge, considering that Wal-Mart estimates its annual case/pallet volume at greater than 8 billion units.

big enough to pick up a signal. The antenna allows a tag to be read at a distance of 10 feet (3 meters) or more, and through many materials including boxes. Antenna size tends to determine the size of an RFID tag. Once an antenna is attached and the assembly is packaged in a protective laminate, the resulting RFID tag becomes finger size or larger, at least for now.

Figure 2.4 illustrates a typical tag IC circuit design. The low-power circuits handle power conversion, control logic, data storage, data retrieval, and modulated backscatter to send data back to a reader.

→ SEE FIG. 2.4

Figure 2.5 shows examples of tags having different antenna designs optimized for various applications. Antennas can be made of silver,

→ SEE FIG. 2.5

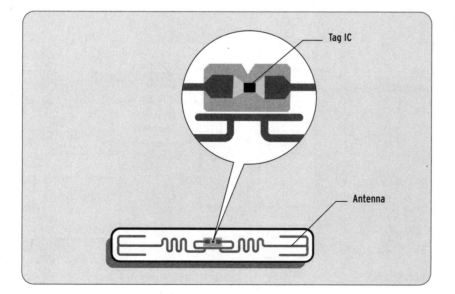

FIGURE 2.3
Tag IC and Antenna.

⊙ SEE PAGE 219 for characteristics affecting read rate.

aluminum and copper, and created by material deposition techniques similar to squirting ink onto a page. The amount of conductive material used and antenna size determines a tag's sensitivity. Sensitivity is crucial, since it has been found that everyday items, such as cases of bottled water or aluminum cans can attenuate RF signals.

Tags are available in bulk quantities in several formats: as plain inlays, as inlays with adhesive backing, as embedded in imprintable labels, or as converted products, where the tag is encapsulated into plastic, rubber or another material and custom designed, molded or laminated.

Tag design, tag placement, case orientation, and reader location all play a role in achieving consistent performance. Tag antennas can

FIGURE 2.4

Typical tag IC circuit design. The low-power circuits handle power conversion, control logic, data storage, data retrieval, and modulated backscatter to send data back to a reader.

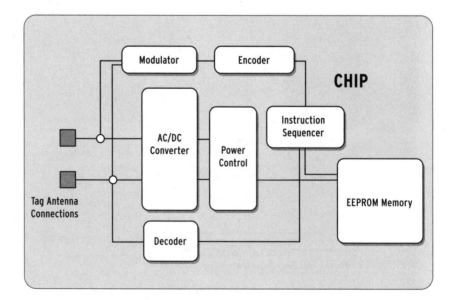

FIGURE 2.5 Examples of various tag designs.

be made in a variety of configurations to achieve various perform-ance characteristics. They need to be an optimal fraction of the operational frequency wavelength. A one-half wavelength is typi-cal. Since the 915 Mhz wavelength is approximately 12.9 inches (33 cm), each one-quarter element wave has an effective length of 3.2 inches (8 cm) long in free space.

Tag antennas are designed to address a broad range of conditions. Dual dipole antennas are less sensitive to physical orientation of the tag to the transmission source than single dipoles. Others are designed specifically for a narrow range of conditions such as for readability when affixed to metal packaging. Tag antennas may also be optimized for reading by a specific type of reader, or with a reader antenna in a specific position.

RFID tag suppliers include:

• Alien Technology, www.alientechnology.com
• Avery Dennison, www.averydennison.com
• Impinj, www.impinj.com
• KSW Microtec AG, www.ksw-microtec.de
• OMRON, www.omron.com
• Philips, www.semiconductors.philips.com
• Rafsec, www.rafsec.com
• Symbol Technology Corp., www.symbol.com
• Texas Instruments, www.ti-rfid.com

Tag development is in its early stages. First generation tags are in current usage, and Generation 2 tags are just being introduced. The number of designs and manufacturers is certain to grow. As standards emerge, and adoption increases, you'll likely find alternate suppliers for the most popular tag types, and lower costs as volumes increase. The expected demand for tags is huge, considering that Wal-Mart estimates its annual case/pallet volume at greater than 8 billion units.

Tag Types

Tags for supply chain use come in a few basic types. One distinguishing characteristic is whether a tag is active or passive. Active RFID tags broadcast under their own power. An on-board battery runs the microchip's circuitry and transmitter. Active tags are capable of receiving and transmitting signals the distance of a football field. They are well-suited to applications where they can be permanently mounted and maintained, such as on railroad cars for tracking movement in a switching yard, on ocean shipping containers, and on high-value military items stored in outdoor supply depots or bases.

→ SEE PAGE 165
DoD use of RFID.

Passive tags, on the other hand, have no battery. Instead, they draw power from the reader. Electromagnetic waves transmitted from the reader induces a current in the tag's antenna. The tag uses that energy to talk back to the reader. The "talk back" is known as backscatter reflection. See Figure 2.6. It is similar to how radar works. Whereas radar backscatter is more like an echo, the tiny circuit in an RFID tag can power itself with the induced current, and its backscatter is a pulse-width modulated (PWM) response. The PWM signal can be interpreted as a digital signal of ones and zeroes.

When they are not in the presence of a reader signal directed at them, passive tags are just that: passive – not capable of emitting any radio signal by themselves. They do not add unnecessary electromagnetic noise to their surroundings.

FIGURE 2.6

Tag-Reader-Host communications.

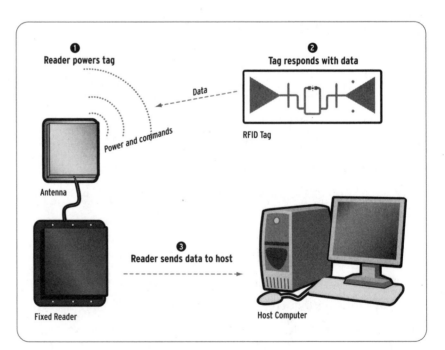

The majority of interest in supply chain RFID is centered on passive tag applications. Tag cost is one key reason. Tag cost is a consideration for supply chain applications even at the case/pallet level. Wal-Mart's estimated case pallet volume is 8 billion alone. A five-cent ($0.05 US) tag cost is widely considered to be the point at which mass adoption will be justifiable (Fig. 2.7). Tag costs are currently several times that number, but are expected to drop rapidly. For many applications, tags have considerable value even at their current cost, and the value will increase as tag usage grows.

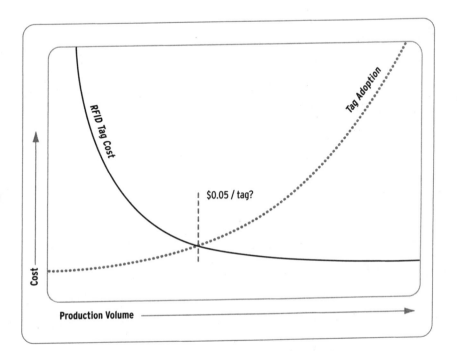

FIGURE 2.7

Tag costs will drop as production volume increases.

Passive tags are smaller in size, lighter in weight, have longer lives and are subject to less regulation than active tags. Passive tags operate only over relatively short ranges, and have limited memory when compared to active tags. Passive tags have difficulty performing in environments where a large amount of interference exists.

Another type of tag is the so-called semi-passive tag. It has many of the characteristics of a passive tag (small, lightweight, limited memory), but with battery backup to extend the answer range. Semi-passive tags are finding uses on shop floor containers and pallets for parts kitting and just-in-time manufacturing applications.

Data Storage Capability

Another major way to distinguish tags is by their data storage capability. Chips in RF tags can be erasable and programmable or read-only. Some first generation tags are programmed by replacing their contents completely. Others have two or more memory pages or blocks that can be programmed separately. The programming range of a tag is in most cases 70 percent or less of its read range. Most applications, therefore, will involve a tag encoder placed in-line with a tag to initially write data into it.

 SEE PAGE 121

on encoding smart labels.

A read-only tag is preprogrammed by the circuit manufacturer. The information on these tags can never be changed. Pre-programmed tag information is linked to a product through a registry entry made in a warehouse management system (WMS) or host computer. Table 2.3 summarizes the basic characteristics of various tag types.

Evolution of Tag Classes

RFID tags have strayed somewhat from the neat classifications originally conceived by the Auto-ID Center. Tag classifications describe both the data and physical layer. Tags that are curently certified for supply chain use are summarized in Table 2.4.

⊕ SEE TABLE 2.4

Tag Type	Advantages	Disadvantages	Application
Active	Greater read range, memory capacity, continuous signal.	Batteries require maintenance. Larger size.	Used with high-value asset tracking.
Semi-passive	Greater read range, longer battery life.	Battery wear and expense.	Reusable containers and asset tracking.
Passive Read/Write	Longer life, multiple form factors, erasable and programmable.	Shorter range than active.	Case & pallet applications. Approved for use with Wal-Mart.
Passive WORM	Suited for item identification, controllable at the packaging source.	Limited to a few re-writes, replacing existing data with new data.	Case & pallet applications. Approved for use with Wal-Mart.
Passive Read Only	Simplest approach.	Identification only, no tracking updates. Difficult to generate on demand and integrate data.	Case & pallet applications. Approved for use with Wal-Mart.

TABLE 2.3

Comparison of active and passive tags with read-only, read/write & WORM.

Class 0 – Passive, UHF based and factory programmed. They are the simplest type of tags. Because the ID numbers in the tags are preset, they are associated arbitrarily to cases and pallets through a host computer at the packaging stage. Wal-Mart has approved a Class 0 version for case/pallet RFID.

Class 0+ – Based on the Class 0 air interface, but programmable (write few, read many). They are accepted by Wal-Mart, DoD and others.

Class 1 – Passive UHF or HF (13.56MHz) based and programmable. The UHF version is accepted by Wal-Mart, DoD and others.

Class 1 Generation 2 – Class 0 and Class 1 tags, which tend to be

TABLE 2.4

Existing tag classes being accepted for supply chain use. See Page 57 for Gen 2 air interface acronyms.

Tag	Class 0	Class 1	ISO 18000-6A	ISO 18000-6B	Gen 2* (ISO 18000-6C)*
Air Interface	PWM-FSK	PWM-PIM	PIE-ASK, FMO	M-ASK, FMO	DSB-ASK, SSB-ASK or PR-ASK
Memory	112 (0), 288 bit (0+)	64, 96 bit	128 bit	96, 256 bit	512 bit
Read/Write	Read only (0), read/write (0+)	Read/Write	Read only	Read/Write	Read/Write
Frequency	902-928	902-928	862-870	860-930	860-960
Security	(0+) Write protect, 24-bit kill	Write protect, 8-bit kill		Write protect	Write protect, 32-bit kill
Suppliers	Symbol/Matrics, Impinj	Alien, Rafsec, Avery	EM Micro	Philips	5 suppliers
*pending					

used in similar supply chain applications, are not currently meeting global performance or interoperable. The next generation that will replace Class 0 and Class 1 tags is UHF Generation 2 Foundation Protocol (Gen 2). It marks the start of a unified passive RFID standard for supply chain applications. The tags that use the Gen 2 standard are expected to be available late 2005.

ISO 18000-6a/b – The Philips UCode 1.19 tag has been accepted by European retail companies, and complies with the ISO 18000-6 specification for passive UHF.

Future tag classes – Class 2 tags are the eventual refinement of the UHF passive tag for supply chain applications, with full data logging capabilities. Class 3 tags are a passive/active hybrid having battery backup, which acts as an internal power source, but it remains in a passive mode until activated by a reader. Class 4 tags are active tags. Each of these class standards has not been fully ratified, and in many cases are evolving rapidly.

Generation 1 Tag Specification

EPCglobal Inc.™ is an independent, not-for-profit, global standards organization entrusted by industry to drive adoption and implementation of the EPCglobal Network™ and EPC technology. RFID specifications cover both the "air interface," how tags communicate, and the programming technique used to store and read data.

By adopting existing designs in the market, most notably the Class 0 from Matrics (now Symbol Technology), and Class 1 from Alien Technology, leaders in the supply chain industry were able to start implementation, prove the value of the technology, and acquire an understanding of how to improve it.

EPCglobal has been supporting initial implementations with a set of standards known as "Gen 1". It is actually a set of specifications:

- **EPC Tag Data Specification Version 1.27** – identifies the specific encoding schemes for trade numbers.

- **900 MHz Class 0 Tag Specification** – the communications interface and protocol for 900 MHz Class 0 operation. It includes the RF and tag requirements and provides operational algorithms to enable communications in this band.

- **13.56 MHz ISM Band Class 1 Radio Frequency (RF) Identification Tag Interface Specification** – defines the communications interface and protocol for 13.56 MHz Class 1 operation.

- **860MHz – 930 MHz Class 1 Tag Specification** – the communications interface and protocol for 860-930 MHz Class 1 operation. It includes the RF and tag requirements to enable communications in this band.

Generation 2 Tag Specification

A new tag standard was ratified by EPCglobal in late 2004, and submitted to ISO for adoption. When adopted, it is expected that the air interface standard will be designated ISO 18000-6c and eventually replace the earlier standards. Gen 2, as it is known, is an attempt to reconcile competing standards and create a harmonized standard that will simplify purchasing decisions for implementers and increase the speed and ease of global adoption.

Gen 2 is the result of work by over forty companies to create a system with cross-vendor compatibility, worldwide interoperability, improvements in performance, reliability and cost. A number of issues have yet to be resolved that may affect Gen 2 adoption. One issue involves patent claims that may require companies building Gen 2 devices to pay royalties. The other obvious issue is availability. A number of companies are currently designing chips, tags, readers and encoders, but production quantities are not expected until late 2005 or early 2006. Wal-Mart and the DoD are saying that Gen 2 will be phased in once available, and tags designed to the earlier specifications will continue to be used over the next two to three-year period.

Will Gen 2 tags and readers provide better functionality, performance and lower costs? Based on the design goals of the specification, and as evidenced in prototype designs and chip emulations, Gen 2 tags should offer these improvements:

➔ SEE PAGE 259
on Gen 2 host to reader
interface.

➔ SEE PAGE 141
for Wal-Mart requirements and Page 170 for
DoD requirements.

Readability – A tighter window in the call and response sequence between a reader and tag, at 4-millionths of a second, will reduce "ghost reads" caused by false triggering of a tag. The reader also has an ability to isolate tags in frequency, and adjustable read rates to accommodate noisy environments. Reader to tag communications can occur in one of up to four sessions. See Figure 2.8. This prevents one reader from interrupting another, for example when a handheld reader transmits within the read range of a dock door portal. With sessions, the portal has one session, and the handheld is automatically assigned a separate session. The tags keep track of each session, and can resume a previous session even when interrupted by a different reader.

FIGURE 2.8

Gen 2 tags can manage multiple sessions with readers.

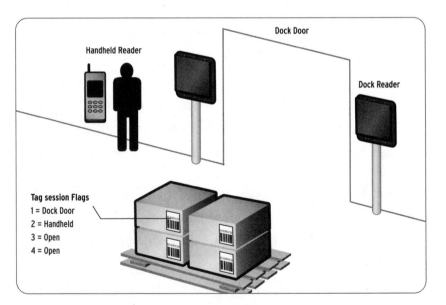

Read speeds – Read speed was addressed in a number of ways. First generation data rates run from 55 to 80 kbps, whereas Gen 2 provides for up to 640 kbps, a ten-fold improvement. This provides a faster singulation rate, the time it takes to identify and isolate a single tag, as well as faster data throughput. Secondly, there is a concept of tag population management where the reader can select and rapidly sort tags by reading only the preamble, or header portion of tags. Faster sorting of tags that respond to a call and on-the-fly adjustments by readers should allow read rates of up to 1000 tags per second. More typically, where there are lots of tags within the reader range, the Gen 2 air interface should provide a 200 tag per second read rate, assuming either the North American or European broadcast environment. As Figure 2.9

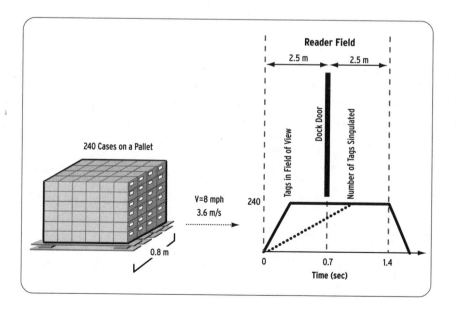

FIGURE 2.9

Read speed can be a factor when tags are moving through a fixed read area.

illustrates, read speed can be an issue if you are required to read all case tags on a pallet as it is moving at normal speed through a portal.

Memory – Gen 2 designs will have double or triple memory capacity. Four distinct memory banks are possible, and each can be addressed separately. See Figure 2.10. Each memory bank can be locked, unlocked, perma-locked and perma-unlocked. The memory banks are always readable regardless of lock status, unless killed or password protected.

FIGURE 2.10

Memory map of a Gen 2 tag.

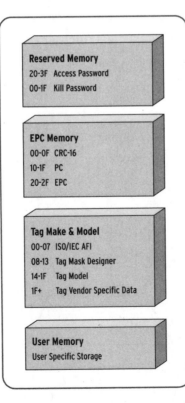

Worldwide operation – The air interface for Gen 2 has improvements that make it more effective within the narrower bands of Europe and Japan, as well as better functionality across the entire 860 to 960 MHz superset of worldwide bands. European RFID deployments tolerate much less interference, and require much tighter spectral control than first generation systems could deliver. This is one reason that Gen 2 has strong backing and support within ISO.

Security – Gen 2 readers have an ability to mask their communications with a tag by first requesting a random number from a tag, then encoding the random number in its tag communications. The tag checks for the number before responding. The random number encoding scheme protects both data and password transactions from snooping. A 32-bit kill password reduces the possibility of someone killing tags in an unauthorized manner.

Cost – The Gen 2 tag IC design lends itself well to very large scale integration. Although volume production and multiple sources of supply, more than anything, will determine costs, Gen 2 chip designs should have a smaller die, so the chips will be smaller than previous generations, which means less material cost. Gen 2 reader chips are based on the same principles as WiFi technology, which should drive reader prices down as well.

Tag Selection and Readability

Product packaging and supply chain processes present a myriad of challenges for RFID. Products and product packages come in all sizes and types, and tags must be able to physically survive the crushing weight of a load. They must withstand shipping wear, temperature extremes, and material handling machinery.

Tag readability is also dependent on characteristics of the UHF spectrum. Here is a list, in no special order, of considerations when

→ SEE PAGE 102
on security enhancements.

selecting a tag for supply chain use:

- **Sensitivity** – The ability of a chip to be energized and send a signal of sufficient strength to a reader.

→ **SEE PAGE 207** on case analysis.

- **Placement and orientation** – Read rate is affected by the orientation of the tag on a box or pallet relative to the reader.

- **Tag position relative to other tags** – Tags can interfere with each other when stacked too close together.

- **Shape and size** – In general, larger tags have a longer ranges. Cases often have a specific place for a label, and some companies specify size and format.

- **Read speed** – The amount of time the tag is within read range in the case of a tagged carton on a moving conveyor, or a tagged pallet on a truck moving through a dock door portal.

- **Read redundancy** – The number of times a given tag can be read while in the reading area. If a tag can respond at least three times to read requests while it is in the reading area, chances are very good that its data will be captured without error.

- **Data requirements** – Tags will contain different information depending on their use (item, case, pallet, returns).

- **RF interference** – Read rates will be affected by sources of RF noise, proximity to other tags, and the composition of packaging materials and surrounding surfaces.

- **Harsh environments** – Steam, corrosive chemicals or extreme cold will affect the adhesive on a tag if nothing else.

- **Re-use** – Re-use could include use on re-usable containers, or as a way to document returned goods.

- **Cross-border shipping regulations** – Tags may have different read range and sensitivity depending on their frequency range of operation, due to different global standards.

- **Collision avoidance** – The number of tags that must be read at once in a given area.

- **Readers** – Available types that support the tag.

- **Progressive use** – Perishable goods, for example, may benefit from a method of logging ambient temperature and expiration.

- **Security** – Some applications may warrant data encryption and other measures that may not be supported in all tag types.

PRINTER/ENCODERS

Passive programmable tags do not have the intended data in them. They require an encoding step to prepare them for use. Encoding can be done by a reader built into an RFID printer, or any reader that is set up for the task. Writing to a tag is more like printing a bar code than it is like reading a tag, even though both are done by an RFID reader (Fig. 2.11). An RFID smart label printer makes an ideal platform for the tag encoding task.

→ SEE PAGE 109
on smart labels.

Table 2.5 contrasts writing to a tag from reading it. When writing data to a tag, a reader has to address a tag individually. The tag must

FIGURE 2.11

Tag encoding in a smart label RFID printer.

be within the proximity of the reader for the time it takes to program it, which may take up to several hundred milliseconds. The tag must be able to draw sufficient power from the reader to enable the programming circuitry in the tag. Isolating the right tag from others around it is very important, to prevent programming the wrong tag. In the case of an RFID printer, tags are encapsulated in a roll of smart labels, and are a known distance apart from one another. Tag isolation is achieved by the design, positioning and tuning of the reader antenna within the printer chassis. The close proximity of the antenna to the tag can be used to advantage, utilizing the properties of near-field electromagnetism to inductively couple to the tag.

→ SEE PAGE 122 on near-field electromagnetic coupling.

	Read	Write
Initial state of tag	Must have data written to it.	Empty or pre-written
Process initiated by	Reader command	Reader command
Tag internal mechanism	Memory poll circuit	EEPROM burn circuit
Response rate	Hundreds per second	Single tag takes a hundred or more milliseconds
Addressing	One to many or one to one	One to one
Sequencing	All tags or listed individuals within read area	Serially and individually
Distance sensitivity	Moderate within effective read range	Extremely sensitive to effective read range
Validation	Multiple reads	Read synchronized with physical isolation of the tag.
Error recovery	Read bar code portion of label	Print overstrike on label and encode next one.

TABLE 2.5

A comparison of tag programming to tag reading.

Closed-loop data validation and error-recovery mechanisms are also built into an RFID printer, making it instrumental to an on-demand tag programming and application process.

Considerations for selecting printer/encoders include:

- **Manual or automated** – A printer/encoder by itself, as a stand alone unit, can be connected to a computer and used to print smart labels for application by hand. Manual operation such as this may be suitable for pallet labeling, cross-docking applications, or piloting a case labeling application with selected SKUs. Automated operation of a printer/encoder involves integrating the unit with an applicator, which automates the process of printing, encoding, verifying, and affixing the label on a case.

- **Labeling requirements** – Your existing bar code and package labeling requirements may or may not accommodate a change to smart labels. A case analysis is required to determine whether or not the location where your existing label is applied is the best location for an RFID tag. A similar size smart label may be available, and a printer/encoder that can print a label that contains all the required labeling information along with an encoded, embedded tag.

- **Data integration** – Your existing bar code labeling data stream, generated by your host computer, will need to be adapted with a command language to trigger a tag encoding operation. Some printer/encoders come with software migration tools that allow the use of your existing software to encode the tag with UPC and GTIN information.

- **System integration** – An automated printer/encoder needs to integrate with the packaging line control system via an input/output module.

- **Environment** – Most manufacturing, warehouse and distribution centers are harsh environments, where equipment is susceptible to physical damage, dust, water, lubricants, temperature extremes, harsh use and electrical noise. Most circumstances call for industrially rugged devices.

- **Dedicated or multi-purpose** – Some printer/encoders are optimized for a certain class of tag, both with firmware and reader type. Others can accommodate Class 0, 0+, 1 and Philips 1.9 (Europe), and accept labels of various sizes and types.

- **Upgradeable** – Printer/encoder manufactures differ in how they provide upgradability. Some companies provide bar code printers that easily upgrade with a customer field kit to encode smart labels, and smart label printers that will upgrade via firmware to accept different types and classes of tags and labels. The eventuality of Gen 2 usage may limit the useful life of a printer/encoder unless the manufacturer has anticipated upgradability in the design of a print platform.

- **Certified** – Since a printer/encoder is in effect a radio transmitter, it must have certifications from the radio spectrum owner (FCC, ETSI, etc.) depending on where it is used. In addition, EPCglobal provides certifications with various tag classes and operating modes.

- **Print speed** – The speed at which a label is printed and the tag is encoded should be considered, especially for production line integrated solutions. Fast moving consumer goods packaging may require multiple synchronized printer/applicators.

- **Print method** – Labels can be produced by direct thermal (treated paper) or thermal transfer (ink ribbon).

- **Precision** – A printer/applicator may need to place labels consistently within a 1/16" tolerance of a location on a case to ensure good reads.

- **Validation and error recovery** – Encoded tags must be validated, and labels should be checked for bar code and quality before being applied to a case.

- **Media detection** – Since many printer/encoders will be used for both RFID encoding and standard bar code printing, it is very important that the printer can distinguish between the two when RFID labels and standard bar code labels are loaded, and can alert the user through an intelligent media detection system when a bar code is about to waste expensive RFID media.

READERS

A reader uses its antenna to send digital information encoded in amplitude or pulse-modulated waveform. A receiver circuit on the tag is able to detect the modulated field, decode the information, and use its own antenna to send a weaker signal response.

Readers are available as handheld devices, mobile mounted (fork-lift or cart), fixed read-only and combination reader/encoder (Figure 2.12). In a typical distribution center, a pair or several pairs of reader/antenna arrays would be configured to identify tags passing between them. Such a configuration is called a portal. Portals may be located at receiving dock doors, packaging lines and shipping dock doors. Mobile mounted and handheld readers can be used to check tags that are not picked up through the portal, or to locate product in the DC or on trucks.

→ SEE FIG. 2.12

Readers have different air interfaces depending on the tag and licensing requirement. Gen 2 readers modulate an RF carrier using double-sideband amplitude shift keying (DSB-ASK), single-sideband amplitude shift keying (SSB-ASK) or phase-reversal amplitude shift keying (PR-ASK) using a pulse-interval encoding (PIE) format. Tags receive their operating energy from this same modulated RF carrier. A reader receives information from a tag by transmitting an unmodulated RF carrier and listening for a backscattered reply.

Because many tags may be in the presence of a reader, they must be able to receive and manage many replies at once, potentially hundreds per second. Tag population management features are used to allow tags to be sorted and individually selected. A reader can tell some tags to wake up and others to go to sleep to suppress chatter. Once a tag is selected, the reader is able to perform a number of

FIGURE 2.12

Typical fixed station and handheld readers.

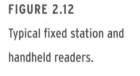

Photo courtesy of Alien Technology, Inc.

Photo courtesy of Applied Wireless Identification Group, Inc.

operations, such as reading the identification number, and writing information to the tag in some cases. The reader then proceeds through the list to gather information from all the tags.

First Generation Basic Operation

Readers serve as gateways between the physical world of tags on packages and the on-line world. Some readers appear as SNMP (simple network management protocol) devices or as internet web servers, with databases that respond to standard SQL (structured query language) commands like other data sources on a network. Multi-protocol or agile readers can identify virtually all classes of tags, or be set to recognize only certain types. Readers that conform to the Gen 2 specification will have variable read speed, and single, multi and dense modes to accommodate various operating environments.

For most applications, readers will operate in one of two ways, either autonomously or as directed/interactive devices. The air interface protocol is the same. They transmit signals in half-duplex, using frequency hopping across approximately 60 bands between 902-928 MHz (USA). Frequency hopping is an FCC requirement to minimize interference with other RF devices. Anti-collision algorithms, combined with a sequence of scroll, quiet and talk commands, are used to read and sort multiple simultaneous incoming tag signals.

Autonomous mode – A reader can be set to continuously operate, accumulating lists of tags in its memory. Tag lists represent a dynamic picture of the current tag population in its read window. See Figure 2.13. As tags respond to reader broadcasts, they are put on the list. If they don't respond they are dropped from the most recent list stored in memory list. A persist time is set to determine the duration between the time a tag was last read and when it is removed from the list. A host system on the network can receive a list of tags from the reader whenever it chooses to listen. The information available to the host would include the reader location, time read, the size of the tag list, and the IDs of the tags on the list.

⊙ SEE PAGE 257

for reader to host
communications.

FIGURE 2.13

A reader set to
read smart labels
autonomously.

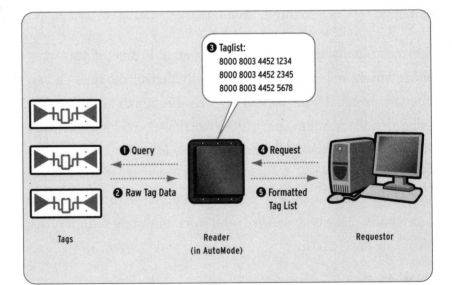

Directed/interactive mode – Readers in this mode will respond to commands from the network host. The host can instruct the reader to gather a list of tags within its read window, or look for a specific tag. In both cases the reader starts by gathering a list. Once it completes the host command, the reader waits until it receives another.

→ SEE PAGE 261 on ALE.

Gen 2 Basic Operation

A Gen 2 reader manages tag populations using three basic operations:

Select – This operation conditionally isolates a group of tags for inventory and access. The select command can use a date code, manufacturer code or other variable that targets only the tags of interest. This operation is analogous to selecting records from a database.

Inventory – The operation of uniquely identifying tags, one at a time. A reader begins an inventory round by transmitting a query command in one of four sessions. One or more tags may reply. The reader resolves conflicts among the tag responses and sets a flag within each of the tags as they are counted.

→ SEE FIG. 10.3 on Page 259 for a conceptual view of a reader.

Access – Once a tag has been identified through the inventory operation, the access operation allows a reader to read from or write to a tag. Access comprises multiple commands to program individual tag memory fields, set passwords, set memory locks or kill the tag.

Considerations for selecting readers include:

- **Operating frequency** – Matched to tag requirements.

- **Multi-protocol** – A desirable characteristic if a variety of tags are to be read which may have different air interface protocols.

- **Encode** – If you expect to use the reader to write data to tags, the command set and encoding capabilities are needed.

- **Meets local regulations** – Power output will be different in the USA and Europe. Frequency hopping is required in the USA and duty cycle in Europe.

- **Memory** – Tag buffering and tag list management features.

⊕ SEE PAGE 252
on network components.

- **Networking capability** – Ability to network readers together, and communicate with host computers through common interfaces (cable, twisted pair or wireless), using RS-485, TCP/IP, Ethernet, or 802.1.

- **Power** – Some readers operate with Power over Ethernet (PoE), eliminating a need for a separate power source, but requiring a PoE router on the network.

- **Configurable and upgradeable** – Through network connection and firmware.

- **Antenna** – Adapts to various conditions using dynamic auto-tuning. Can accept multiple antennas for various applications.

- **Control interfaces** – Digital input/output and control circuits for synchronization with other components on an automated line.

- **Gen 2 upgradeable** – As Gen 2 tags become available, readers may need a firmware upgrade at a minimum.

READER ANTENNAS

Reader antennas are the most sensitive component of an RFID system. Typical antenna set-ups for packaging line and dock door reading are shown in Figure 2.14 and 2.15. Most reader antennas are housed in enclosures that are easy to mount, and tend to look like plain, shallow boxes. Varying the reader antenna placement is usually the easiest adjustment to make when troubleshooting a system, and one of the trickiest things to do well. The reader antenna must be placed in a position where powering the tag and receiving data can be optimized. Since government regulations limit the broadcast power of a reader, antenna placement is vital to achieving a high read rate.

→ SEE FIG. 2.14
→ SEE FIG. 2.15

→ SEE PAGE 214
on reader antenna placement.

Figure 2.16 shows a portal designed to detect directionality. This configuration might be used at a doorway between the stockroom

→ SEE FIG. 2.16

FIGURE 2.14

Antenna configuration for conveyor "curtain".

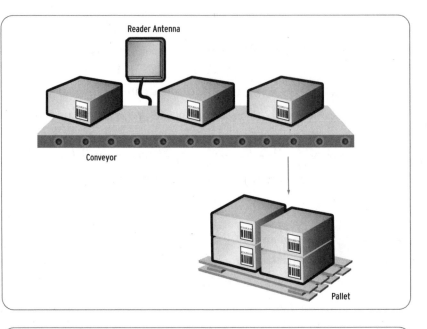

FIGURE 2.15

Antenna configuration for a dock door "portal".

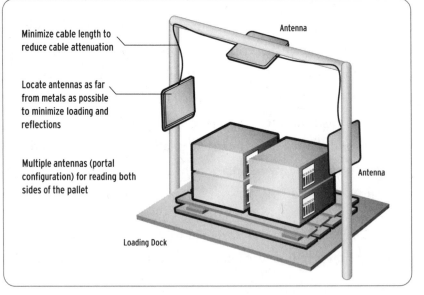

and the retail floor. Forklifts carrying RFID tagged cases going through the doorway will be read by antennas on one side before the other. Software in the inventory system then notes the two reads within a close time window, and interprets it as a direction. This allows the system to note that cases were carried from the stockroom to the retail floor. When the cases are detected going back to the stockroom, the system interprets that the cases have been emptied. This might be confirmed by another portal at the trash compactor.

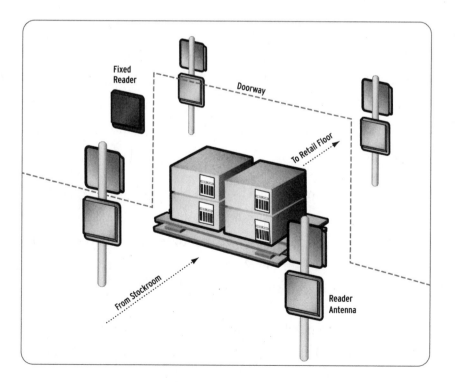

FIGURE 2.16

Portal antenna array designed to detect movement and directionality.

Three characteristics of antennas contribute to tag readability:

- **Pattern or footprint** – The three-dimensional energy field created by the antenna. This is also known as the reading area.

- **Power and attenuation** – The maximum power of a reader antenna is fixed in order to meet FCC and other regulatory requirements. The signal can be decreased or attenuated, however, to limit the tag read window or aim it only at tags you want to read.

- **Polarization** – The orientation of the transmitted electromagnetic field.

Linear antennas typically provide the longest range, but are sensitive to the orientation of a tag (Fig. 2.17). They can be used in instances such as an RF reading curtain mounted over a conveyor. The tags would be affixed to packages in a consistent orientation to maximize readability.

Circular polarization is created by an antenna designed to radiate RF energy in many directions simultaneously (Fig. 2.18). The antenna offers greater tolerance to various tag orientations, and a better ability to bounce off of and bypass obstructions. These abilities come somewhat at the expense of range and focus.

UHF antennas are nearly always externally mounted and connected to a reader via shielded and impedance matched coaxial cable.

Linear Polarized Reader Antennas:

RF energy radiates from antenna in a linear pattern

The wave has a single E-field component

Generally longer range than a circularly polarized antenna when tag is optimally oriented

Can have a narrower beam pattern than a circularly polarized antenna

Best for applications with known tag orientation

FIGURE 2.17

Linear polarization.

Circular Polarized Reader Antennas:

RF energy radiates from antenna in a circular pattern

The two E-field components are equal in magnitude, 90 degrees out of phase and spatially oriented at 90 degrees from one another

Designed to increase signal reception in presence of multipath and high scattering

Offers more tag orientation insensitivity, slightly reduced range from linearly polarized antenna

FIGURE 2.18

Circular polarization.

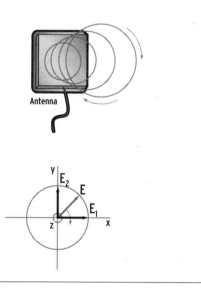

One or more antennas can be connected to a signal reader, depending on your application requirements. An antenna is first selected for the operating frequency and application (omni-directional, directional, etc.). De-tuning, or signal weakening can occur due to the following:

- RF variations
- Skin-effects
- Losses due to metal proximity
- Antenna cabling losses
- Signal fading
- Proximity of other reader antennas
- Environmental variations
- Harmonic effects
- Interference from other RF sources
- Eddy Fields
- Signal reflections
- Cross talk

Some of these effects can be compensated through dynamic auto-tuning, circuits in the reader, which work with feedback from an antenna's resonance tuning parameters. In most cases, antenna placement is not pure science, and on-site adjustments are required to achieve optimal read rates.

INTEROPERABILITY TESTING

In order to provide guidance to end users, and encourage the adoption of these standards by device manufacturers, EPCglobal periodically tests various readers and printer/encoders. The results are published on their web site (www.epcglobalinc.org). The tests show that solutions are available from device suppliers environments where a mix of tags are having to be read and used. In addition, a not-for-profit testing facility called the RFID Alliance Lab is testing and publishing its findings on tag read range and other response characteristics.

An EPCglobal Gen 2 Testing and Certification workgroup has been chartered to:

• Review Gen 2 Certification Test Plans for RF and Protocol Testing currently being developed by MET Labs and its partner, CETECOM Spain.

• Provide technical feedback on UHF Gen 2 Certification Test Plans via comment matrices.

• Actively resolve all technical feedback through comment resolution process.

EPCglobal will begin performing UHF Gen 2 conformance and interoperability testing and will publish their first results at the EPC Symposium in September 2005.

THE GEN 2 STORY

by John Schroeter, with Dr. Chris Diorio, Impinj Inc.

The recently ratified EPCglobal Gen 2 RFID standard—the new state of the art in supply chain technology—has raised the bar for competitive RFID systems worldwide. Unfolding in tandem with recent well-publicized retailer and DoD mandates, it would seem that the timing of the new standard couldn't be more opportune. But like the promise of a great western frontier, getting there might require the crossing of a few uncharted mountains and deserts.

While the theory behind practical, widespread RFID deployment is sound enough—perhaps even proven—one would still do well to regard the words of Yogi Berra: "In theory, there's no difference between theory and practice. In practice there is." Impinj's Dr. Chris Diorio, a significant contributor to the Gen 2 spec development effort, agrees. "A spec is a framework on which you build a solution, not a solution in and of itself. Even though the Gen 2 spec represents a huge technological advancement compared with current Gen 1 RFID systems, and offers a solid foundation upon which to build an RFID system, fielding a working Gen 2 system is still a big step beyond the spec."

Deployment hurdles notwithstanding, Gen 2 offers a worldwide standard for RFID systems operating in the UHF band. A remarkable achievement, it was crafted with the cooperation of more than forty participating companies and driven by a shared vision of cross-vendor compatibility, worldwide interoperability, and significant improvements in performance, cost, and reliability over predecessor UHF protocols. Furthermore, it offers a roadmap that extends well into future generations, as future UHF standards will be built upon the Gen 2 foundation. All of which tills the ground for fertile investment in a technology that will redefine a critical element of commerce: the supply chain. The specification is so widely endorsed that it is now under the wing of the venerable ISO organization, where it is slated for early 2006 ratification as ISO 18000-6C.

Presently deployed Gen 1 UHF RFID systems are based on a number of competing protocols, most notably Class 0 and Class 1. The problem? The current incarnations of these protocols lack the features, reliability, and horsepower to adequately serve a growing number of applications—particularly when taking worldwide operability into account. And because they're not upgradeable, their shortcomings, in many cases, mean that they have already hit the wall.

MIT's Auto-ID Center recognized the problems of these proprietary RFID standards early on. More importantly, they also recognized that provincial protocols would impede the development and large-scale deployment of RFID technology. Their solution? A single open standard that 1) would create an environment of interoperability and international regulatory compliance, and 2) would raise the bar on RFID system performance in a significant way. These two values formed the backbone of what they proposed as the next generation of UHF RFID—the precursor to the Gen 2 standard. With a single worldwide specification in place, UHF RFID-based systems would become faster, easier to use and less costly to field, more robust, and provide a multi-supplier path going forward.

The Auto-ID Center "kicked-off" the Gen 2 effort in June of 2003, at a seminal meeting in Zurich, Switzerland. They would eventually transfer the responsibility for development and commercialization of the evolving standard to EPCglobal, who, in December of 2004, finally ratified the standard as "Generation-2 UHF RFID Protocol for Communications at 860 MHz – 960 MHz." Or more simply, Gen 2.

At the center of this activity was University of Washington Professor, Dr. Chris Diorio. Working with the renowned Dr. Carver Mead—the father of digital VLSI—Diorio had previously developed a low-power nonvolatile memory (NVM) technology that would make possible long-range, field-rewritable RFID tags. This work caught the attention of MIT Professor Sanjay Sarma, the Auto-ID Center's Chairman of Research, and in turn led Diorio to join the Auto-ID Center. With Dr. Diorio's background in communications, semiconductors, and standards work, he was soon appointed co-chairman of the Hardware Action Group and project editor for the Gen 2 specification, roles that would continue with Auto-ID's technology transfer to EPCglobal.

In the early stages, Gen 2 faced significant resistance, as it became apparent that it would obsolete all of the existing UHF RFID standards. But Diorio, the end users, and dozens of technology companies persevered, convinced that Gen 2 held the greatest promise for proliferating UHF RFID. Says Diorio, "Gen 2 is the first worldwide RFID standard that's backed not only by the technology providers, but by the end users as well." The momentum behind Gen 2 continued to build, and eventually all parties—including those who were initially opposed to the effort—joined in the final push to ratification.

The spirit of innovation that led to 802.11 wireless radios, and to cellular phones, is very much in evidence throughout the pages of the Gen 2 specification, the objective of which was to bring harmony out of the cacophony of competing and otherwise incompatible standards. With that in mind, consider these twelve Gen 2 qualities that serve to position the new standard at RFID's center stage:

1. Superior Tag Population Management If the Gen 2 development effort had a mission statement, its cornerstone would have been the efficient and accurate management of RFID tag populations. To that end, Gen 2 defines the interactions between readers and tags over a robust air interface with three primary command-driven procedures: Select, Inventory, and Access. Let's take a closer look.

Select – Prior to conducting an inventory, a user may wish to first conditionally isolate only those tags that exhibit, say, a particular date code, manufacturer code, or other variable of interest. By targeting only that segment of the tag EPC memory that contains those particular kinds of descriptive bits, the reader can quickly narrow down the field, making for a more efficient inventory operation. The Select command offers a quick sorting of the tag population, where the reader (using union, intersection, and negation operators on a set of user-defined selection criteria) chooses a subpopulation of the tags within its field.

Inventory – The Inventory operation—the real workhorse of the Gen 2 protocol—identifies tags, one at a time, resolving conflicts among tag responses, and sets an appropriate "inventoried" flag within each of the tags as they're counted.

Access – Available only following an Inventory operation, Access involves more than simply sorting and counting tags; Access commands allow the reader to write individual tag memory fields directly (with EPC and/or password data), set the desired memory lock bits, or kill the tag.

2. Robust Signaling Protocol Consider the way a Gen 2 reader uniquely identifies a single tag within a population. When a reader issues a Query command, the tag must respond within an extremely narrow window—just 4-millionths of a second wide. If the tag does not respond within that timeframe, the reader assumes that no tag is present, and issues another Query command. The reader continues to poll in this manner until it receives a valid response. This tight window represents the first hurdle in a series of "communication qualifiers" designed to eliminate false triggering on noise and other spurious emissions. When a tag does respond, it does so with a preamble—a distinctive waveform that the reader is able to reliably discern and identify, even in noisy environments. If the reader does not recognize the preamble as the leading part of the tag's response, it is ignored.

As data begins to flow from tag to reader in the form of well-defined symbols, memory retained in the waveform is used to identify bad sequences, or alternatively, to make decisions on ambiguous bits and fix them. Diorio illustrates with an example. "If you were to hear the phrase, 'a stitch in time saves pine,' you would know that, based on the history of that set of words, that it should read, 'a stitch in time saves nine.' In the same way, through the signaling scheme described in Gen 2, information contained in sequential symbols allows the reader to correct errors as it goes along."

Once the transmission is complete, the reader reviews the waveform and checks the PC (Protocol Control) bits at the top of the transmission, used to compare the number of bits it received with the number of bits the tag says it sent. If the two numbers match, then the reader can be fairly confident of a valid transmission. But not so fast. The reader then compares the CRC (cyclic redundancy check) at the end of the transmission and verifies its integrity. Only then is the reader satisfied that it has read a valid EPC.

Says Diorio, "True Gen 2 readers, implementing these checks properly, will have an astronomically low probability of seeing ghost reads. In fact, with a properly implemented Gen 2 system, end users should simply not tolerate ghost reads."

3. Dense-Reader Operation Any truly practical vision of RFID deployment will require the fielding of many readers, all of which might be operating simultaneously, blasting away at full volume, and in fairly close proximity to one another. Faint-voiced tags will have little, if any, hope of being heard above such a din; the noise and interference resulting from of all that shouting will surely bury them. Unless they're Gen 2 tags.

Gen 2 gets around the problem of "dense" readers by isolating tags and readers through a frequency channelization scheme, best illustrated by analogy:

Think of the UHF spectrum as a highway, where readers are semi trucks, and tags are bicycles. Dense-reader spectral planning effectively divides the UHF frequency band—the highway—into multiple lanes. Trucks are allowed to use certain lanes, while bicycles are permitted to use other lanes. In any case, trucks and bicycles do not share lanes. Furthermore, trucks are required to remain within the boundaries of their respective lanes, as sideswiping or drifting into the bicycle lanes would prove disastrous (particularly for the bicycles!). In similar manner, if a reader's signal (which is many orders of magnitude greater than that of the tags) were to leak into adjacent tag lanes, it would mask the tags' low-power transmissions, burying them in RF noise, and preventing other readers from seeing them at all. By restricting reader transmissions to occur within strictly delineated lanes (or channels), tags can be heard clearly, even though as many as 50 active readers might be operating simultaneously in 50 available channels.

The fact is there is a limited amount of spectrum allocated for UHF RFID (this is particularly true in Europe). A Gen 2 system incorporating only dense-reader capable readers is best able to exploit that space efficiently, and it does so to great effect.

4. Cover-Coding of the Forward Link Maintaining a secure link between reader and tag is essential to safeguard data transmitted over an air interface. It's especially critical in the reader-to-tag direction, because reader transmissions occur at substantially higher power levels than those of the tags, who effectively whisper their responses back to readers. Here's how Gen 2 goes about the enciphering:

The reader requests a random number from the tag. The reader then mixes that random number with its data before transmitting the result to the tag. The tag decodes the mixing (reversing the operation) and extracts the original information. A simple scheme, but it effectively protects both data and password transactions by obscuring data transmissions in a purely random manner.

5. A 32-bit Kill Password Incorporated to address privacy concerns, the kill command permanently disables a tag from talking back to a reader, rendering it useless. The ability to kill a tag, though, exposes the network to the possibility of unauthorized kills. To thwart such mischief, a password protection scheme was adopted. In earlier UHF RFID versions, though, it didn't amount to much. Class 1's 8-bit kill password, for example, left it exposed to only 256 possibilities—hardly a password at all. While Class 0 improved things significantly with a 24-bit password, Gen 2 raised the hacker's bar to 32 bits—more than 4,000,000,000 possibilities.

6. Sessions Dramatically Boost Productivity Capitalizing on the enabling power of the dense-reader mode, Gen 2 also introduces the concept of sessions, where as many as four different readers may access the same population of tags through a time-interleaved process. That's an extremely useful capability. Consider the case where a shelf-mounted reader in the midst of a counting operation (assigned to, say, session 1), is interrupted by another reader entering the field—possibly a handheld reader—to perform its own inventory operation (in session 2, perhaps). Dock door and forklift readers, assigned to sessions 3 and 4 respectively, might also jump in for a round. Because Gen 2 tags maintain a separate "inventoried" flag to keep track of each of these various random and independent sessions, they're able to seamlessly resume their participation in the previous (pre-interruption) inventory round, picking up right where they left off, and never miss a beat.

7. Significantly Faster Singulation Rates Readers and their associated tag populations have a lot of business to transact. At least one throughput-gating parameter is the effective data rate, which also determines the time it takes to singulate, or identify, a single tag within a population. Typical Gen 1 data rates run from 55 to 80 kbps. Pretty fast. Gen 2, though, provides for data rates as high as 640 kbps (a throughput of 1600 tags per second). That's on the order of a tenfold performance improvement over Gen 1.

8. Variable Read Rates The rate of data flow between reader and tag is governed by a number of factors: environmental conditions (including noise level and physical structures), region of operation, the number of active readers in the area, and even the speed of tagged materials moving through the distribution center. A very adaptable system, Gen 2 allows the fine-tuning of the RFID network—including the varying of data rates—to optimize performance across all possible combinations of operating conditions. Whether the need is for fast reads of pallets moving through a dock door, or slower reads in a noisy environment of dense readers, Gen 2 is able to flex.

9. Greater Configuration Control The modulation and data encoding schemes of choice, like the selection of data rates, also depend on a set of environmental considerations. As such, Gen 2 provides options in both reader-to-tag link and tag-to-reader link directions, allowing performance calibration of the Gen 2 system to the demands of its operating environment. Gen 1 systems, on the other hand, are limited to a fixed communication format, where one size may not necessarily fit all.

10. Worldwide Operation Both Gen 1 and Gen 2 systems cover the 860 to 960 MHz operational band—the superset of international frequencies—but the way the two standards deal with that spectrum is worlds apart. As far as Europe is concerned, Gen 1 doesn't deal with it very well. Compared with North America's fairly wide frequency allocation of 902 to 928 MHz, Europe's is pretty slim—just 865 to 868 MHz. As such, European RFID deployments tolerate much less interference, and require much tighter spectral control than Gen 1 systems can deliver. Gen 2, on the other hand, takes the European standards fully into account; it works well in North America, Japan, Europe, and elsewhere, making Gen 2 a truly international standard, hence its strong advocacy within ISO.

11. Improved Manufacturability Gen 2 tag IC design lends itself well to very large scale integration. The new standard exhibits a greater emphasis on digital logic; as Moore's law continues to shrink fabrication geometries, Gen 2 tags will get ever smaller still.

12. The whole is greater than the sum of the parts. While each of the foregoing features represents a significant performance improvement over Gen 1 offerings, when taken together, they leverage much more. Further, as Gen 2 systems are fielded, it will become apparent that perhaps the greatest lever of all is the communications systems engineering expertise behind the specific implementations. This is where the Gen 2 standard is forged into proven elements that comprise a working system solution.

Sources and Further Reference

Computer Networks, Third Edition, Andrew S. Tanenbaum, Prentice Hall, 1996.

A Basic Introduction to RFID Technology and Its Use in the Supply Chain, white paper, by Steve Lewis, Laran RFID, January 2004.

RFID Primer, white paper, Alien Technology, 2002.

RFID for the Supply Chain: Just the Basics, white paper, Printronix, February 2004.

Understanding the Wal-Mart Initiative, white paper by John Rommel, Symbol Technology Corp., 2004.

The Gen 2 Story: Charting the Path to RFID that Just Works, white paper by Impinj, Inc., 2004, available: www.impinj.com.

Hardware Interoperability Testing Results, available: www.eglobalinc.org.

EPC™ Radio-Frequency Identity Protocols Class-1 Generation-2 UHF RFID Protocol for Communications at 860 MHz – 960 MHz Version 1.0.8, 2004, EPCglobal.

CHAPTER 3

From UPC to EPC

STANDARDS DEVELOPMENT THROUGH EPCGLOBAL   85

ELECTRONIC PRODUCT CODE   87

EPC Format   88

EPC Representations of Standard Identity Types   90

EPC & RFID COMPARED TO UPC AND BAR CODES   92

TAG SECURITY   100

Security Enhancements in Gen 2 Specification   102

ADDRESSING CONSUMER CONCERNS ABOUT RFID   106

IN THIS CHAPTER:

Details on the Electronic Product Code. How it compares to bar code.

Bar codes have been with us for over fifty years. The first bar code patent was granted in 1952. It has only been in the last 30 years, however, that bar codes became ubiquitous. Mass adoption followed three converging mechanisms – industry mandates to use them, standardization, and refinements of labeling and scanning technology.

Like bar codes were 30 years ago, RFID is in very early adoption. See Figure 3.1. Unlike bar codes, RFID technology and standardization has not matured in advance of retail mandates for adoption. A

FIGURE 3.1

Bar codes gained mass adoption over a 30-year period.

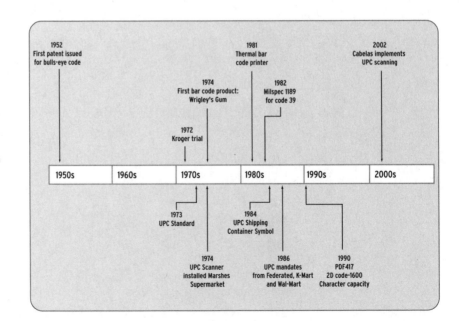

dozen years went by between when UPC bar codes were standardized and retail giants mandated their use.

At this point we cannot say whether some of the more visionary proclamations about RFID (clothing tags that tell the laundry machine how to wash them, for example) will ever happen, or if RFID will ever completely replace other forms of machine identification. Mandates, standardization and technology refinements can set us on a path; it's anybody's guess where that path will lead.

→ SEE PAGE 137 on industry initiatives.

STANDARDS DEVELOPMENT THROUGH EPCGLOBAL

Standardization gained focus through the Department of Mechanical Engineering at MIT, with the formation of the Auto-ID Center in October 1999. The Auto-ID center championed passive RF tag designs and manufacturing techniques to drive down costs. In order to keep tags as simple as possible, memory capacity was limited to a few thousand characters. An information architecture was drawn up, where tags serve as keys or pointers, wirelessly linking items to information stored in databases accessible via the internet.

→ SEE PAGE 268 for global exchange of supply chain data.

This led to the idea of the Electronic Product Code, or EPC, a comprehensive key linking an item, case or pallet to detailed information anywhere in the supply chain. The goal, however, was not to replace bar codes. Rather, the goal was to create a migration path for companies to move from bar code to RFID. The administrative functions of the Auto-ID Center officially ended on October 31, 2003 and the research functions evolved into Auto-ID Labs. EPCglobal works very closely with Auto-ID Labs to refine the technology and meet future needs. EPCglobal Inc. is a joint venture of GS1 (formerly EAN International) and GS1 US™ (formerly Uniform Code Council) which administers the UPC bar code. GS1 represents 101 member organizations worldwide. The EAN.UCC System – the world's most accepted standards system – is used by more than 260,000 companies doing business in over 140 countries.

Table 3.1 lists some of the standards organization involved in one way or another with RFID. Although differences exist among these organizations, a breathtaking amount of consensus and standards making has occurred, especially considering the short time involved. With EPCglobal taking the lead, the Gen 2 specification effort involved over 60 RFID companies worldwide. The consensus is enough to make several major retail companies and the US Defense Department feel comfortable with the risks and rewards, and mandate adoption by their supply chains. The roadmap for the standards making effort calls for getting Gen 2 ratification from ISO by 2006, at which time it will become ISO18000-6c.

ELECTRONIC PRODUCT CODE

Like the UPC, the EPC is divided into numbers that identify the manufacturer, product serial number and version. The EPC ranges from 64 to 256 bits, with four distinct fields as shown in Figure 3.2. What sets the EPC apart from UPC is its serial number, which allows individual item tracking.

→ SEE FIG. 3.2

Organization	Mission	Website	Domain
ANSI	Commercial trade standards	www.ansi.org	USA
APEC	Commercial trade standards	www.apec.org	Asia-Pacific
APICS	Supply chain business process	www.apics.org	Global
Auto ID Center	RFID standards	www.epcglobalinc.org	Mission continued by EPCglobal
CEN	Commercial trade standards	www.cenorm.be	Europe
ETSI	RF broadcast standards	www.etsi.org	Europe
EAN	Supply chain ID & EDI	www.ean-int.org	Global
ECCC	Bar codes and e-commerce	www.eccc.org	Canada
EPCglobal	EPC & RFID	www.epcglobalinc.org	Global
FCC	RF broadcast regulations	www.fcc.gov	USA
GS1 US	Supply chain	www.gs1us.org	USA
ISO	Commercial trade standards	www.iso.ch	Global
MPHPT	RF broadcast standards	www.soumu.go.jp	Japan

TABLE 3.1

Standards organizations involved in RFID.

EPC Format

The EPC format is an open format capable of describing physical entities for a number of purposes, including supply chain RFID tag applications. The format can also be used in bar coding and other machine-readable encoding applications.

Note that the EPC number is not the only data stored in a tag. Standard UPC bar code information can be encoded to a tag, for instance. The US Department of Defense has another classification system for suppliers who may not be members of EPCglobal. The EPCglobal Gen 2 data specification calls for a numbering system identifier (NSI) which supports the ISO Application Family Identifier (AFI). The NSI will distinguish tags used for EPC supply chain use from other tags on PCs, vehicles or other objects that happen to be within the range of a reader.

The general format for EPC tag data includes these sections:

Header – The header identifies the version number of the code itself.

FIGURE 3.2

Electronic Product
Code format.

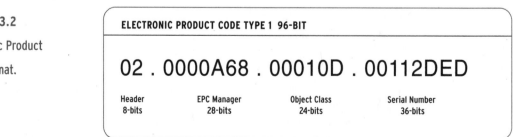

ELECTRONIC PRODUCT CODE TYPE 1 96-BIT

02 . 0000A68 . 00010D . 00112DED

| Header | EPC Manager | Object Class | Serial Number |
| 8-bits | 28-bits | 24-bits | 36-bits |

EPC Manager – Identifies an organizational entity (e.g., a company, a city government) that is responsible for maintaining the numbers in subsequent fields – Object Class and Serial Number. EPCglobal assigns the General Manager Number to an entity, and ensures that each General Manager Number is unique.

Object Class – Refers to the exact type of product, similar to a SKU (stock keeping unit) The object class is used by an EPC managing entity to identify trade items. These object class numbers, of course, must be unique within each General Manager Number domain. Examples of Object Classes could include SKUs of consumer-packaged goods, or road signs, lighting poles or other highway structures where the managing entity is a county.

Serial Number – A unique identifier for the item within each object class. The managing entity is responsible for assigning unique, non-repeating serial numbers for every instance within each object class.

The EPC Type 1 number, at 96 bits in length, will accommodate as many as 268 million companies, each having 16 million classes, with 68 billion serial numbers in each class. In Class 1 tags, an additional 32 bits of the EPC are for unique item information (item description, ultimate destination, special handling instructions, etc.) that can be reused at any point in the supply chain.

The 96 bit EPC is more than enough to serialize every product world wide for years to come.

EPC Representations of Standard Identity Types

EPC versions of six identity types are described in the EAN.UCC tag standard working draft (Fig. 3.3). These identity types were developed as bar code standards, and the tag standard proposes EPC coding schemes for them.

Global Trade Item Number (GTIN®) – This is the formal term for a globally unique EAN.UCC System identification number for products and services. A GTIN may be 8, 12, 13 or 14 digits. To create a unique identifier for individual items, the GTIN is augmented with a serial number, which the managing entity is responsible for assigning

FIGURE 3.3

Standard Identity types and what they represent.

GTIN
Global Trade Item Number

GRAI
Global Returnable Asset Identifier

SSCC
Serial Shipping Container Code

GIAI
Global Individual Asset Identifier

GLN
Global Location Number

UID
Unique Identification

uniquely to individual object classes. The combination of GTIN and a unique serial number is called a Serialized GTIN (SGTIN).

Serial Shipping Container Code (SSCC) – The SSCC is assigned uniquely by the managing entity to a specific shipping unit. A section of the code is set to quickly distinguish basic logistic types such as a pallet from a case.

Global Location Number (GLN) – GLN can represent functional, physical or legal entities. Examples of functional entities are a purchasing or accounts payable department. Physical entities might include a particular warehouse room or loading dock. Legal entities can include whole companies, subsidiaries and divisions.

Global Returnable Asset Identifier (GRAI) – Like a GTIN, except the object class is used to identify the asset type. Examples of returnable assets include barrels, pallets, gas cylinders, beer kegs, rail cars and trailers.

Global Individual Asset Identifier (GIAI) – GIAI is used by a company to label fixed inventory or any property used to carry on the business of the company. Examples are hospital beds, computers and delivery vehicles.

Unique Identification (UID) – UID is the DoD asset tracking number, which is being harmonized into the EPC standard.

Other classifications – As EPCglobal evolves the classification system and global registries to embrace other industries and applications, we can expect standardized encodings for the vehicle identification number (VIN) used in the transportation industry, the commercial and government entity (CAGE) used in the US federal procurement system, and others.

EPC AND RFID COMPARED TO UPC AND BAR CODES

For the past 25 years bar codes have been the primary means of identifying products in the supply chain. Bar codes have been effective, but have limitations. The key attributes to consider when comparing RFID and bar coding center around reading capability, reading speed, tag or label durability, amount of information, flexibility of information, cost and standards. A migration toward RFID involves a number of considerations, one of which is whether it should complement or replace bar coding for the application.

Read Method – Seeing and hearing are two different things as they say (Fig. 3.4). Line of sight has its advantages. Bar code optical readers offer an absolute visual verification. The reader signals a good read within its aim, and a bad read is immediately associated with a specific label and item. This is a one-to-one relationship. RFID is more like hearing. It does not require line-of-sight to read the tag information. The radio frequency (RF) signal is capable of traveling through most materials. This is particularly advantageous in warehouse receiving operations, and in operations where information needs to be collected from items that may have an inconsistent orientation, such as distribution center sorting applications. An RFID reader is able to distinguish and interface with an individual tag despite multiple tags that may be within the given read range. The discrimination of tags, however, does not include with it the absolute physical location identification that a bar code reader has when it is aimed at a specific point on a packaging line. Tags that do not respond (quiet tags) for one reason or another require a manual search and verify step, or diversion of

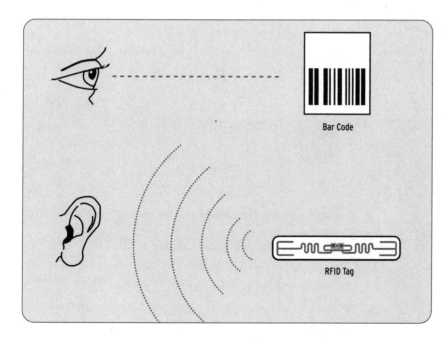

Bar Code

RFID Tag

FIGURE 3.4
Roughly speaking, a bar code is like seeing and RFID like hearing.

the entire pallet for further investigation. Suddenly, a whole pallet is held up. A batch process now requires batch level recovery mechanisms.

Read Speed – RFID tags can be read far more rapidly than a bar code label, at theoretical rates of 1,000 per second (Fig. 3.5) or more. This far surpasses the one-at-a-time reading speed of bar code. The speed of RFID has great value in high-volume receiving and shipping applications where a large number of items need to be accounted for quickly. For example, when receiving a pallet of tagged cases in a warehouse, an RFID reader can potentially identify all cases without having to break the pallet down and scan each

FIGURE 3.5

Comparison of reading speeds of a bar code versus RFID.

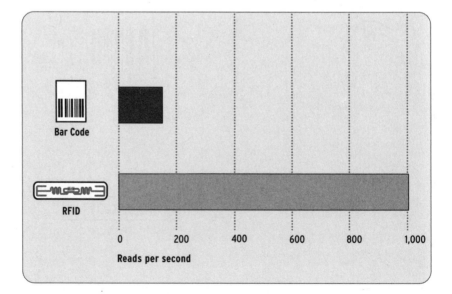

individually. Electronics manufacturers use bar codes during the assembly of printer and computers. Hewlett-Packard scans bar codes on tote trays 256 times in the course of making a product. An RFID tag could reduce misreads and speed up parts identification.

Readability – Read rates approaching 100 percent are possible with bar codes in high-speed automated lines engineered for bar code identification. The engineering practice is established and results are repeatable. RFID offers the promise of perhaps better read rates at equal or greater line speed, but the engineering practice is in its early stage. Achieving a high accuracy goal, such as Wal-Mart's 100 percent mandate for RFID, may be non-trivial for many companies. Figure 3.6 summarizes some of the components of read accuracy.

→ SEE PAGE 219 on characteristics affecting read rate.

COMPONENTS OF READ-RATE:

Bar Code	RFID	
Environment	Reader Antenna	Tag Antenna
Line speed	Environment	Case content
Label to scanner orientation and distance	Line speed	Encoded and validated
Meets symbology specifications for contrast and legibility	Protocol	Label Positions

FIGURE 3.6

100 percent readability will involve process, environmental and technical engineering.

Durability – RFID tags can be encased in hardened plastic substrates or other materials. Although they are significantly more durable than paper bar code labels, both depend on adhesive to hold them intact and attached to an item. Bar codes etched on metal or plastic have proven reasonably durable over the years. RFID has been used to track engine blocks throughout their production process, which is often too harsh for a bar code. Further, the durable nature of RFID tags allows them to last longer than bar code labels. The Achilles' heal of an RFID tag is the mating point of the antenna to the chip. A cut that severs through the mating point will disable the tag, whereas as a bar code may be only slightly degraded.

Data Storage – UPC identifies an item classification, but EPC can identify an individual item through an assigned serial number. The traditional linear or 1D bar code stores up to 100 characters, and a 2D bar code can store 1,000 characters or more. See Figure 3.7. High-end RFID tags may contain several kilobytes of memory (several thousand characters). This increased information storage capability creates a portable database of information, allowing a greater number of product attributes to be tracked, such as date of manufacture, time spent in transit, location of distribution center holding the item, expiration date or last date of service. On the other hand, a typical bar code shipping label contains an enormous amount of information in human readable and bar code form, including addresses for shipping, content markings, means of conveyance, etc.

A smart label combines both storage capabilities.

 SEE PAGE 227
on smart labeling
approaches.

Flexibility of Information – With respect to information dynamics, RFID tags are able to support read/write operations, enabling real-time information updates as an item moves throughout the supply chain. The total information model for passive tag RFID, however, does not depend on re-writing the tag. The EPC number written on a tag serves as a key to information stored in external databases. These databases support the concept of recording the history of a tagged item throughout its lifecycle. See Figure 3.8. This feature can be of critical importance as production schedules, delivery dates and locations, and shipment contents can change on

→ SEE PAGE 277
on using RFID in the
supply chain.

→ SEE FIG. 3.8

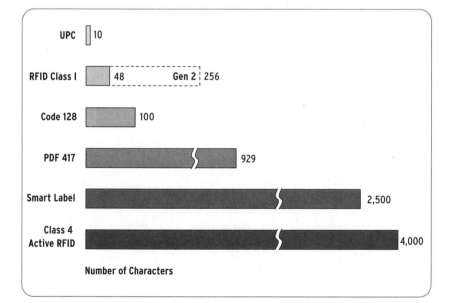

FIGURE 3.7
Comparison of data
storage capacities.

a regular basis. Bar codes, of course, can also be used to record an EPC number, and serve as keys to external databases. In most supply chain applications, however, they are not used this way.

Information Redundancy – RFID tags hold information in captive form, offering it only through a reader tuned to accept it. System integrity is non-linear – you either accept or reject what the reader says. Bar codes, on the other hand, usually have a human-readable character format adjacent to the bar (Fig. 3.9). This permits a direct recovery should a bar code fail to read. Think of how many times you have avoided a long wait at the grocery checkout counter as the clerk picks up the package and keys in the bar code number after it

FIGURE 3.8

RF tags have potential over the entire product life cycle.

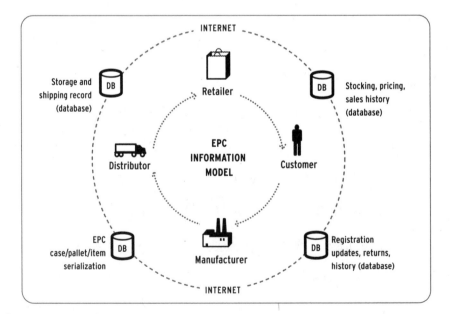

failed to scan. Smart labels, containing bar code and human readable characters along with the RFID tag, may offer the best combination of information redundancy and integrity.

Security – Both RFID tags and bar codes have encryption routines that support various information security requirements. Two dimensional matrix bar codes are impossible for a human to read, just as with RFID tags. Some tags, however, support a password scheme that can render them unreadable to reading systems that do not use the password to access the EPC code. Security is treated in more detail on the following page.

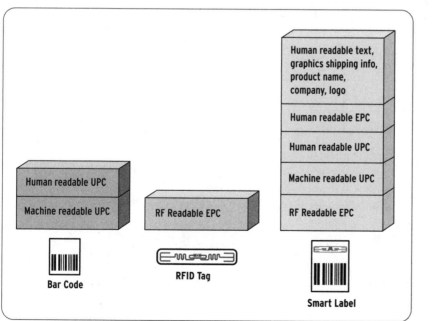

FIGURE 3.9

Comparison of information redundancy.

Cost – RFID requires new capital and operating investments. Figure 3.10 shows the typical investments and costs. A return on investment (ROI) cost calculation may justify RFID only when the risk of not doing so means losing a prime customer, such as Wal-Mart, Target or DoD. A pragmatic approach, where the investment level keeps you in business as you gain knowledge and experience, may be the least cost approach.

TAG SECURITY

Among the many popular misconceptions about RFID is that they are more vulnerable to information theft than other identification methods. Can tags be "hacked?" Can someone in an unauthorized manner read a tag and perhaps replace its memory contents? Of course, but it gets harder as air interface protocols improve. Can readers be "spoofed," by a counterfeit tag? Probably.

FIGURE 3.10

Typical startup costs for a retail distribution center RFID implementation.

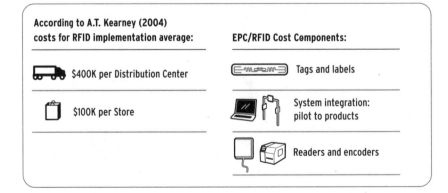

According to A.T. Kearney (2004) costs for RFID implementation average:

$400K per Distribution Center

$100K per Store

EPC/RFID Cost Components:

Tags and labels

System integration: pilot to products

Readers and encoders

In 2004, Lukas Grunwald, a consultant at DN-Systems Enterprise Solutions GmbH in Germany, using published tag air protocol specifications, was able to read and rewrite RFID data on tags at a Metro store. Grunwald wanted to demonstrate what he feared would be a way for shoplifters to alter the data on an expensive item to read as though it were a commodity. The circumstances, however, involved the pilot use of a read/write tag that lacked the safeguards of a true item-level system.

The original Auto-ID Center vision for RFID was for tags to contain "license plate" – type data, serving as pointers to secure databases that contain the real information about the item. As tag technology evolves and on-board memory increases, more theft sensitive information could potentially be stored on the tag.

→ SEE PAGE 261 on ALE.

What current tags contain, should you read one, is less than what you can read off of a bar code shipping label. A tag EPC provides the unique item classification, the source identification, and a serial number. Pricing and other sensitive information will be stored on external databases, with well-established security. The EPC can be locked with a password. Tags can also be set to quiet mode, and won't respond without a password command. Table 3.2 lists the basic security measures that are part of the command sets of current and Gen 2 tags.

→ SEE PAGE 267 on security and quality of service.

→ SEE TABLE 3.2

Security Enhancements in Gen 2 Specification

Enhanced security is one improvement made with Gen 2 tags. By comparing Gen 2 security measures to an established network security model, we can get an appreciation of how secure an RFID system could be. ISO 7498 is the security specification used to evaluate network security architectures. It defines five main categories of security service:

- **Authentication** – Typically used at the start of a connection, authentication checks a claimed identity at a point in time.

- **Access control** – Protection against the unauthorized use of a resource for reading, writing or deletion of data.

- **Data confidentiality** – Protection against the unauthorized disclosure of information.

- **Data integrity** – Protection against active threats to the validity of data.

TABLE 3.2

Basic security features in tag command sets.

Feature	Description	Tag Types
Lock	Write protection that prevents reprogramming a tag, using a password	Class 0+, 1, UCode 1.19 and Gen 2
Encryption	Encodes the EPC	Class 0+, 1 and Gen 2
Kill	Renders a tag completely inoperable	Class 0+, 1 and Gen 2

• **Non-repudiation** – Protection against the sender or recipient of data denying that the data was sent.

The Gen 2 specification describes commands and protocol between a reader and a tag population, and between a reader and a host. Both layers of communication can be wireless, potentially, and vulnerable to security threat. Obviously, there is minimal risk to an organization if a single tag is compromised, and the risk escalates if an intrusion compromises a tag population, an event manager, a warehouse database, a supply chain execution system, and so on. EPCglobal has enlisted the help of VeriSign, Inc. to help implement security measures for information sharing between trading partners and within the EPCglobal Network of registry and discovery services.

⊙ SEE PAGE 257

for Gen 2 host to reader interface.

In a fully implemented security model for a Gen 2 RFID system, each layer has a number of mechanisms that deliver one or more of the five security services defined in ISO 7498. See Table 3.3.

⊙ SEE TABLE 3.3

• **Reader to tag authentication** – Each of four memory blocks in a tag can be assigned a 32-bit password which a reader must give to authenticate itself and initiate a session.

⊙ SEE PAGE 48

for a Gen 2 memory map.

• **Reader to tag access control** – The password is required to initiate various tag commands to select, read, write, delete, lock or kill require a password. A 32-bit password protection translates to 4 billion possibilities, making it much more difficult to create mischief. In addition, data written to a tag can be perma-locked to ensure it cannot be changed or written over.

⊛ SEE PAGE 77

on cover-coding.

• **Reader to tag data confidentiality** – To ensure confidentiality, a reader cover-codes its transmission by first requesting a 16-bit random number from a tag. The reader mixes the random number, or handle, with the data transmitted to the tag. The tag reverses the mixing, and looks for its handle within each reader transmission during a session. Cover-coding helps maintain a confidential one-to-one link, preventing the interception of passwords and data broadcast from a reader.

• **Reader to tag data integrity** – Signal redundancy in the waveform, protocol control, plus a 16-bit CRC ensures integrity. As data flows from a tag to a reader, memory retained in the waveform is used to identify and correct ambiguous parts of the transmission. When complete, the reader reviews the waveform and checks the protocol control bits at the top of the transmission

TABLE 3.3

Gen 2 enhancements support a robust security model. At higher levels, a number of established corporate computer security mechanisms can be implemented to limit the risk.

	Reader to Tag	Host to Reader	Host to Host
Authentication	32-bit password	WiFi Protected Access (WPA), Kerberos, 802.11, etc.	VPN, Firewall, SSL, etc.
Access Control	32-bit password	User defined	User defined
Data Confidentiality	Cover-coding	User defined	User defined
Data Integrity	PC, CRC	User defined	User defined
Non-repudiation	ACK	User defined	User defined

with the number bits the tag says it sent. Then it verifies the CRC at the end. In addition, a reader can compare tag data with a record on a host computer database to validate the data. Gen 2 tags also have session flags to keep track of conversations with more than one reader. This prevents one reader from interrupting another and causing the tag to lose track of an exchange.

◉ SEE PAGE 46
on Gen 2 readability.

• **Reader to tag non-repudiation** – An acknowledgement command is used to effect non-repudiation. In addition, a narrow call-response time-out window of 4 ms prevents ghost-reads, and triggers a repeat exchange.

The security measures for supply chain RFID may have vulnerabilities, but they are an improvement over bar codes. Anyone with a laser printer can scan a bar code on a low-cost item, print out copies and stick them on a higher priced item, and use a self-checkout lane at a retail outlet to get away with paying less. Spoofing a tag reader is more clandestine than counterfeiting a bar code label, but safeguards at the reader and payment system level will more easily detect this kind of fraud. The payment system can compare the EPC against its inventory and determine whether it is valid or cloned or fictitious.

When viewed as a total closed-loop system, it is clear that RFID will improve automated detection and loss prevention within the supply chain. At the pallet, case and eventually the item level, RFID will

→ SEE PAGE 156
on pharmaceutical
initiatives.

reduce "shrinkage," which is the term for losses from poor stock keeping and employee theft. EPC validity checking is being mandated for pharmaceutical track and trace systems to prevent drug counterfeiting. For retail merchandise, the EPC will also automatically register the buyer for warranty, rebates and returns eligibility, providing security and other benefits to the buyer, who is now the owner.

ADDRESSING CONSUMER CONCERNS ABOUT RFID

Although RFID implementations are currently focused on supply chain case and pallet tracking, the long-term goals of the retail industry include the tagging of items. Early trials and media stories about them have raised public concerns about RFID, and public advocacy against its use. The concerns relate to the privacy of individuals as they acquire and use products with embedded RFID tags.

The retail industry realizes that public acceptance cannot be taken for granted. In response, EPCglobal has adopted guidelines for the use of EPC on consumer products. These guidelines are certain to evolve, and currently address these areas of concern:

Consumer Notice – Consumers will be given clear notice of the presence of EPC on products or their packaging. This notice will be given through the use of an EPC logo or identifier on the products or packaging.

Consumer Choice – Consumers will be informed of the choices that are available to discard or remove or in the future disable EPC tags from the products they acquire. It is anticipated that for most products, the EPC tags would be part of disposable packaging or would be otherwise discardable. EPCglobal, among other supporters of the technology, is committed to finding additional efficient, cost effective and reliable alternatives to further enable customer choice.

Consumer Education – Consumers will have the opportunity easily to obtain accurate information about EPC and its applications, as well as information about advances in the technology. Companies using EPC tags at the consumer level will cooperate in appropriate ways to familiarise consumers with the EPC logo and to help consumers understand the technology and its benefits. EPCglobal would also act as a forum for both companies and consumers to learn of and address any uses of EPC technology in a manner inconsistent with these Guidelines.

Record Use, Retention and Security – The Electronic Product Code does not contain, collect or store any personally identifiable information. As with conventional technology, data which is associated with EPC will be collected, used, maintained, stored and protected by EPCglobal member companies in compliance with applicable laws. Companies will publish, in compliance with all applicable laws, information on their policies regarding the retention, use and protection of any personally identifiable information associated with EPC use.

Sources and Further Reference:

The Bar Code Book, by Roger C. Palmer, Helmers Publishing, Peterborough NH, 1989.

RFID Explained, A Basic Overview, white paper, Robert W. Baird & Company, February 2004.

A Basic Introduction to RFID Technology and Its Use in the Supply Chain, white paper, by Steve Lewis, Laran RFID, January 2004.

Guidelines on EPC for Consumer Products, available: www.epcglobalinc.org.

"RFID's Security Challenge," by George Hulme, Thomas Claburn, *Information Week,* November 15, 2004.

The Gen 2 Story: Charting the Path to RFID that Just Works, white paper by Impinj, Inc., 2004, available: www.impinj.com.

EPC™ Radio-Frequency Identity Protocols Class-1 Generation-2 UHF RFID Protocol for Communications at 860 MHz – 960 MHz Version 1.0.9, 2004, available: www.epcglobalinc.org.

From Bar Codes to Smart Labels

ANATOMY OF A SMART LABEL 112

SELECTING THE RIGHT SMART LABEL FOR THE JOB 116

LABEL CERTIFICATION 119

ENCODING, PRINTING AND VALIDATING SMART LABELS 121

TROUBLESHOOTING TAG READING PROBLEMS 127

SMART LABELS COMPARED TO OTHER APPROACHES 128

IMPLEMENTING SMART LABELS 131

DEPENDENCIES AND POINTS OF CONTROL 131

ENTERPRISE-WIDE SMART LABEL PRINTING MANAGEMENT 134

IN THIS CHAPTER:
How smart labels combine RFID with bar code for case/pallet pilot applications.

Supply chain labels provide product identification and logistical information to support shipping and handling of product from manufacturing to point of sale. Smart labels are supply chain labels with embedded RFID tags. They offer promise in helping organizations deploy RFID for compliance with retail industry and DoD mandates. Smart labels allow you to retain bar code/shipping label information in the same or similar format to what you are currently using, while adding RFID. In effect, the embedded RFID tag becomes a wireless bar code.

Smart labels provide a convenient and economical way to package RFID tags and stream them into the distribution process. They provide a richer data set than either a tag or bar code by themselves. They provide resiliency, by combining bar code and human-readable text together with electronic data, should one method of identification fail. They comply with consumer and industry guidelines for having a visible indication that a package has an RFID tag. They can be produced on-demand, or pre-printed and pre-coded for batch processing. Labels also provide added protection to the tag from heat, dust and humidity.

⊙ **SEE PAGE 227**
on smart labeling approaches.

Smart labels may be the easiest, least disruptive, least costly way to implement RFID in your facility. They are not only appropriate for case and pallet supply chain applications, they may also be used in a number of applications "within the 4 walls," including receiving, routing, stocking, work-in-process, HAZMAT and asset handling.

Smart labels offer a convenient method for tagging a product or asset. If you are already producing bar code labels, a migration to smart labels could involve the integration and re-use of an established process, where:

- On-demand printing and application flexibility is maintained
- Labeling is done at appropriate points in the packaging/shipping process
- RFID integration fits within the small footprint of a smart label printer
- Both automated and operator-assisted application methods are available
- Tag encoding is done predictably and reliably, without custom engineering
- Validation and error recovery is built into the system
- Encoding and printing commands share an established host computer to shop floor network
- System migration and integration can be simplified using conversion tools and software modules from multiple middleware and supply chain execution system suppliers, so you don't have to rewrite applications.

ANATOMY OF A SMART LABEL

Figure 4.1 provides two views of a smart label. The surface area is used for standard bar code and label text. The RFID tag is sandwiched in the middle. The label sandwich typically consists of seven parts – the liner or carrier sheet, liner release coating, inlay adhesive, tag inlay, facestock adhesive, label material and a label topcoat.

Label converter companies package RFID tags into various sizes and types of labels depending on the application. For example, the "squiggle antenna" and "wave antenna" (Figures 4.2a and b) are designed for general purpose applications. The "four-T antenna"

FIGURE 4.1

Two views of a smart label.

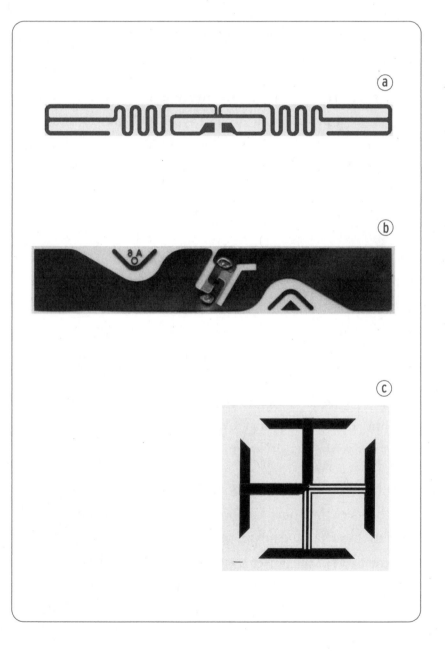

FIGURE 4.2

Smart labels with tags for various uses.

SEE FIG. 4.2C

(Figure 4.2c) is a dual-dipole antenna designed to function even when packages are rotated in various orientations.

As the smart label market grows over the next few years, you are likely to see a proliferation of label types and sources of supply. Many converters are beginning to source RFID chips independently, create their own antenna designs and develop new substrates and other label components to improve readability, durability and differentiate their product.

Label converters are hoping to deliver high yields and low prices in the pursuit of market share. Smart label cost elements are shown in Figure 4.3. Since production volume tends to dramatically reduce costs, prices for smart labels could actually go up over the next year or so, before they come down. This is due to the high up-front investment costs converters have to pay to get into the market, the number of vendors having to share a relatively small market near term, the anticipated change-over to Generation 2 tags, and the lack of a "one size fits all" antenna and label design.

Smart labels are not just like other labels produced by converters. They contain electronic components that can fail due to mis-assembly, mechanical damage, mishandling and ESD. Tags that fail to operate within the power level parameters of the UHF air interface specification, for whatever reason, are called quiet tags. Quiet tags will continue to be a source of discussion as the industry refines its standards for tag quality and testing.

To meet the demand for 100 percent read rates in the market, converters at this time can only offer yield guarantees at the time of shipment. They inspect tags to determine the exact yield. They then remove and replace defective labels, or ship extra quantities of labels to compensate for a less then 100 percent yield coming out of the manufacturing process. Some printer/encoder companies take this concern away for the user by offering quality guarantees for specific labels that are certified for their printer/encoders.

⊕ **SEE PAGE 141**

for Wal-Mart system

requirements.

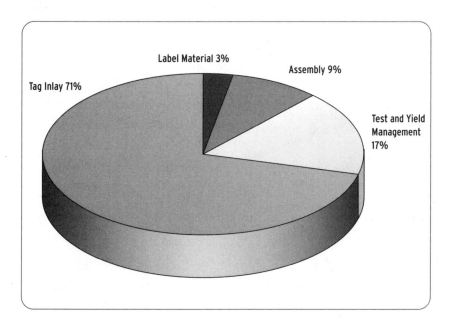

FIGURE 4.3

Cost components in the

label converting process.

SELECTING THE RIGHT SMART LABEL FOR THE JOB

Smart labels are available in bulk roll or fan fold quantities in a variety and sizes and types. You'll find formats that allow smart labels to substitute for existing bar code labels designed to meet corporate and supply chain standards (mil spec labels, freezer grade labels, pharmaceutical chain of custody labels, etc.). Three basic criteria should drive selection of the right label:

Labeling requirements – The information on the surface of the label and embedded in the tag needs to identify the package contents and its status in the supply chain. Supply chain data could include originator, shipper, ship to and any special handling requirements. Figure 4.4 shows a typical label format, with address areas and machine readable codes. The amount of information on the front of this label typically exceeds what is stored in EPC format on the RFID tag. EPC codes, after all, were originally designed as "license plate numbers" that reference an external database record with detailed information on the product.

⊖ **SEE PAGE 207**

on case analysis.

Application requirements – Tag readability is a major factor. A complete case analysis is required to ensure tag readability. Case analysis should include a comprehensive assessment of package contents, packaging design, label placement and the package labeling process. Package contents and packaging design may affect tag readability, particularly if metals, liquids, high carbon or salt

FIGURE 4.4 Typical format for case labeling.

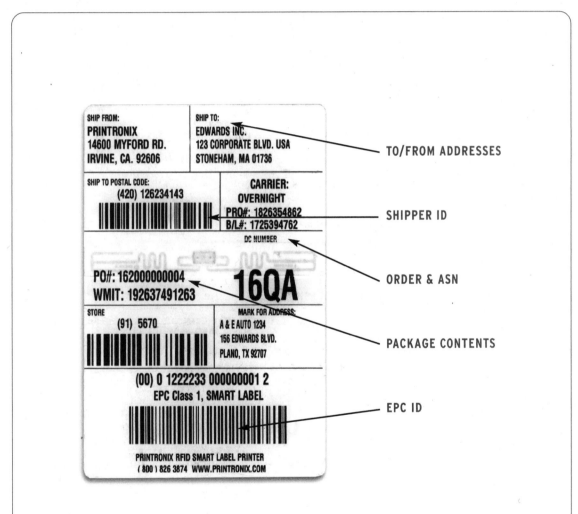

content is involved. This affects label placement on the package, choice of label, and package orientation with respect to readers as it moves through the process. Storage and handling requirements determine whether you need such things as freezer-grade adhesive, or polyester rather than paper label material to withstand heat. One method for collecting data for the case analysis is to set up a "basketball court" with the reader at the basket and a tagged case moved around the three-point circle to assess readability at various distances and angles.

LESSONS LEARNED

Customer requirements – Customer specifications may dictate everything from the volume needed, to the size of label used, the information printed on label surface, location of the label on a package, the EPC information in the tag, the tag/reader air protocol, and acceptance criteria. Acceptance criteria may include such things as yield, advanced shipping notice (ASN) integration, validations, record keeping and problem handling (charge-backs). The stringency of your customer's acceptance criteria will greatly affect the design criticality of your process. As your labeling process moves into production, the number of labels needed and line speed or throughput will dictate the level of investment needed to meet a customer requirement.

→ SEE PAGE 142
on ASN integration.

Clearly, selecting where to put an RFID tag or smart label on a product is not as easy as applying a bar code label. The engineering trial and error involved is one of the biggest challenges facing companies who have to meet a RFID compliance mandate. Application requirements, as defined by a case analysis, usually determine the choice of smart label. Case analysis will match the application to a specific tag, antenna design, and placement location.

LABEL CERTIFICATION

Although smart labels are available from a number of sources, certain labels may be incompatible with certain printer/encoders. The process of inserting RFID tag inlays into labels is new for many label converter companies. They have encountered many challenges, particularly ESD management, inlay placement accuracy and inlay testing.

Reasons for incompatibility include:

- Tag type not matched to the encoder. (Tags are categorized as class 0, 0+ or 1 and have incompatibilities. The Gen 2 standard will eventually resolve this.)
- Incompatibility between the tag antenna and printer/encoder antenna causing poor near-field coupling.
- Tag not at a position within the label so that it is correctly aligned with the printer/encoder antenna.
- Label surface material needs to be suitable for high quality transfer or direct thermal printing.

• Adhesive needs to be matched to the package surface to which it will be applied.

High speed package labeling requires a high quality label. Labels that are difficult to peel, or inconsistently peel from their backing because of poor die cuts will fail and require operator intervention. Poor adhesive application during the label converting process can result in labels lifting as they travel through a printer, causing a jam. The core diameter of label rolls need to be matched to label sizes to reduce the amount of induced curl on labels as the roll unwinds to its core. Curl or adhesion problems will affect how well the vacuum system on an applicator can hold a label in place. Label carrier sheet stiffness can also contribute to applicator problems by making it difficult for the label to separate as it passes through the peel bar.

Certified labels have been pre-tested to eliminate incompatibility, and ensure optimum performance. Label certification is especially important with RFID, since the programming of the tag and synchronization of tag data with printed label data occurs at the printer, all within a second's time. A one percent error rate on labels, for example, when production throughput is 40 labels/minute, would result in almost 200 rejects a day.

ENCODING, PRINTING AND VALIDATING SMART LABELS

Labels are printed most often using a thermal transfer print process. A print head, embedded with a matrix of tiny heat elements that are precisely controlled, uses heat to transfer a wax or resin ink to the label. The ink is on a separate ribbon inside the printer. Smart labels come in rolls of various sizes which, along with the ribbon roll, are mounted inside the printer/encoder. An alternative to the label/ribbon system is direct thermal media, whereby chemicals in the label turn black when heated by the print head.

Initially, most passive UHF tags have no data in them. They require an encoding step to load the data. Encoding can be done by a reader built into an RFID printer, or any reader that is set up for the task. Note that the word "reader" is a general term for a device that can both write to (encode) and read from RFID tags. A smart label printer makes an ideal platform for the tag encoding task for a number of reasons:

Singulation – When reading tags, the reader starts by compiling a list of tags, which it can poll individually if it wants. When writing data to a tag, a reader has to address a tag individually. Isolating the right tag from others around it is very important, to prevent programming the wrong tag. Blank tags won't respond to a call. Some tags have null data or another code in them, inserted by the tag manufacturer during final testing. A reader calling to those tags may get the same response from all of them. The only way to synchronize the reader with a blank or null-data tag is by uniquely positioning the tag within a precise read window. That way, the reader has a tag, one tag only, and the right tag within its grasp. A printer/encoder performs this task within a controlled environment.

Proximity – Within approximately one wavelength ($\lambda/2\pi$) from a power transmission source, conductive materials experience the affects of near-field electromagnetism. See Figure 4.5. Within the near-field, magnetic coupling occurs. Power flows in the direction of the magnetic field. Near-field energy transfer is more like that of an electric motor, or a power transformer. At greater than one wavelength, radio waves separate and propagate away, no longer aligned with the magnetic field. Energy drops off by the square of the distance. At 900 MHz, the near-field breaks down within a few

FIGURE 4.5

Near-field and far field electromagnetism.

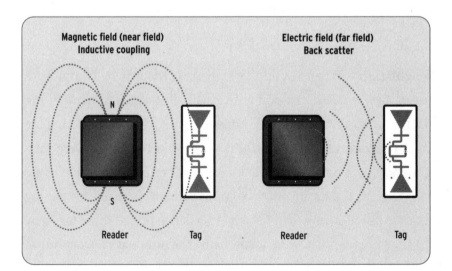

inches of the source, therefore a reader antenna more than a foot away from a tag cannot take advantage of near-field magnetic coupling. Within a printer/encoder, the tag to encoder antenna position is optimized for near-field energy transfer.

Power and duration – Compared to a read command, a write command requires a higher power level and longer duration. The tag must be able to draw sufficient power from the reader to drive the programming circuitry in the tag. The tag must be within the proximity of the reader for the entire time it takes to program it. A printer/encoder provides this.

Tag singulation can be a challenge, since many tags can potentially respond to a reader. In the case of an RFID printer, tags are encapsulated in a roll of smart labels, and are a known distance apart from one another. Singulation is achieved by the design, positioning and tuning of the reader antenna within the printer chassis. The close proximity of the antenna to the tag is used to advantage, by utilizing the properties of near-field electromagnetism to inductively couple the tag. Because of the encoding time needed, label production time is somewhat less than it takes to produce a bar code label only. The time trade-off, however, is more than compensated for by precise process control, high duty cycles, and validation and error recovery routines that eliminate bad tags and labels being applied to a package.

⊙ **SEE PAGE 227** on print and apply approaches.

Figure 4.6 shows the interior side view of a printer/encoder. The sequence of operation is as follows:

1. The printer receives its commands from the host computer. Once a label is in position, the reader does a pre-check. Some tags have null data encoded by the tag manufacturer as a part of its parameter testing system. The null data helps distinguish good tags from quiet tags that may have escaped detection during the smart label converting process. If a tag is quiet during the pre-check, the printer ejects it by marking it with a strike-over (Fig. 4.7).

→ SEE FIG. 4.7

2. A single command by the reader programs the tag. The command includes an erase, write, read back and verify sequence, along with the specific EPC data to be written to the tag. Tags cannot be partially written to. A write command replaces data that might be in the tag. If the read back and verify is not successful, the command is repeated.

3. The reader then performs an explicit read back validation process, to again match the tag data reply and price response level with what it was expecting. If verification cannot be confirmed the label is overstruck with a pre-defined print pattern and ejected from the printer.

4. Labels with verified tags are printed using the bar code and text character job stream associated with the EPC number.

FIGURE 4.6 Side view of the inside of a smart label printer.

5. If the printer has ODV (Online Data Validation) installed, it also reads the bar code information with a bar code scanner mounted in-line with the label exit path. ODV compares the bar code with ANSI standards for dimensional tolerance, edge roughness, spots, voids, reflectance, quiet zones and encodation. If the label does not pass, the printer provides an option to reverse its path, overstriking the label to indicate that it failed the test, and ejects the label.

FIGURE 4.7

Examples of good and bad (quiet) labels.

A label is considered good when the RFID data is written to the tag correctly, the correct image is printed, and content data is verified agaiinst the source.

If the printed and encoded data can't be verified against the source, the label is considered defective can be voided from the system.

6. A record of the production sequence is sent to the host computer.

7. If the printer is integrated with an applicator or other packaging line components, it will communicate with logic controllers to sequence next steps in the production line.

TROUBLESHOOTING TAG READING PROBLEMS

A properly encoded tag produced by a smart label printer is one sure step toward having 100 percent readability on cases and pallets. Should your process not result in a full yield of properly reading labels, some good troubleshooting work is necessary. Some areas to investigate and eliminate as suspect include:

- Tag and reader compatibility
- Read range and speed of movement of the label through the read area
- Tag to reader orientation
- EMI emissions in the environment
- Interference from case contents
- Damage during label application
- Placement of label on packaging consistent with case analysis

Other considerations include using the data log at the printer/encoder to verify that a good read took place as it exited the printer. Determine if there is a discrepency between the yield specification supplied by the label converter, and the log of final test results from the printer/encoder.

SMART LABELS COMPARED TO OTHER APPROACHES

There are alternatives to smart labeling for RFID, and the choices are expected to increase to accommodate item tagging. Let's look at the pros and cons of other approaches:

Tags integrated with packaging – Disposable corrugated packaging with built-in RFID has several attractive characteristics, most notably that it eliminates tag handling during packing and sealing. Encoding can occur before or after packing. Tag acquisition and integration costs are pushed down to your suppliers. Disadvantages may include cost and the implications of managing both RFID and non-RFID packaging. Other drawbacks include a lack of clear marking for RF, lack of a backup identification of what has been programmed into the tag, and a huge potential for rework costs because once the box is built, tag location can't be changed. Industry analysts predict that it will take 3 or 4 years for packaging companies to overcome the physical hurdles and make available a comprehensive offering. Error recovery, rework, and charge-backs may actually drive up costs.

Pallets and totes with permanent RFID tags – Such pallets offer a simple approach to pallet level identification, especially where the tags are programmable, and the pallets are dedicated to a specific route in the supply chain. Similarly, for direct-to-store shipments, totes with fixed tags may gain acceptability. This approach might suit the tagging of work-in-process goods during manufacturing as well. Disadvantages when compared to smart labels include the difficulty in encoding unique EPCs with each use, and re-purposing totes and pallet loads. Also, many totes and pallets do not return to point of origin. The expected extended life of such tags require ruggedized packaging.

RFID tags only – Another possible approach is applying adhesive backed inlays directly to cases and then programming them when they travel through the packaging line. They might be used in conjunction with an existing shipping label. Although tag-only application may appear to be a less-costly, simpler method, there are a number of drawbacks. Tag inlays, by themselves, are currently not available with the broad selection of adhesives that exists today for labels. RFID tags by themselves do not provide the human readable and bar code backup that smart labels provide. They are subject to the same issues of adhesion to moist, frozen or non-flat surfaces as smart tags, but without being as obvious when they fall off. Consumer privacy groups have also requested a clear indication when an RFID tag is present, which a smart label provides. In addition, smart label printers have validation/recovery routines that prevent a bad tag from ever being applied to a case.

← SEE PAGE 106

on addressing consumer concerns.

Applying just the tag, then programming it while it's moving down the packaging line, necessities package-level rework should the tag fail to read. Lastly, placing individual tags on previously labeled cartons could result in a mismatch between the bar code labels and the RFID tags.

Table 4.1 summarizes the advantages and disadvantages of various approaches compared to smart labels. The observations made here are in the context of pilot applications and initial deployments with the retail industry (Wal-Mart et al) and Department of Defense mandates over the 2 years. In the future we can expect to see a variety of approaches to integrating RFID with today's high-speed retail packaging environments.

TABLE 4.1

Comparison of various tagging approaches to smart labels.

	Boxes with built in tags	Pallets/totes with permanent tags	Tags directly applied to cases	Smart labels
Tag placement	Fixed and pre-engineered	Fixed and pre-engineered	Flexible	Flexible
Tag application	Built-in	Built-in	Applicator	By hand or by applicator
Encoding sequence	Programmed on packaging line	Programmed at shipping dock	Applied then programmed	Programmed then applied
Activation	Curtain	Portal	Curtain	Printer/encoder device
Error recovery sequence	Case unpack & repack	Pallet unpack & repack	Rework	Detection & recovery before applied to case

IMPLEMENTING SMART LABELS

Smart labels can play a key role in an RFID migration strategy. An externally applied adhesive smart label is the easiest, quickest, most flexible way to go from "no tag" to "tag selectively," paving the way to "tag everything." They bridge laboratory and pilot applications by putting tags into play in various environments, so that tag placement, tag orientation, read range, read rates, reader placement, and data management issues can be identified and resolved. In most cases they provide a production or near production line solution at least cost. In all cases they offer a backup reading capability to aid troubleshooting and recovery.

⊙ SEE PAGE 179
on getting started.

Smart labels are a ready-fit format for most RFID implementations. EPC data flows down to them through a host computer system and printer/encoder. See Figure 4.8. Labels are applied to parts, products, packages and pallets, uniquely identifying them. Smart-labeled objects are now linked up by radio frequency to a supply chain execution system.

⊙ SEE FIG. 4.8

DEPENDENCIES AND POINTS OF CONTROL

Whether the smart labeling system is minimally designed for initial compliance requirements or has a high level of integration with other systems, a number of areas need to be managed.

Critical dependencies and points of control include:

EPC numbering – Case, pallet and eventually item serial numbers need to be created, encoded in EPC format, and recorded somewhere

FIGURE 4.8

Operational data flow.

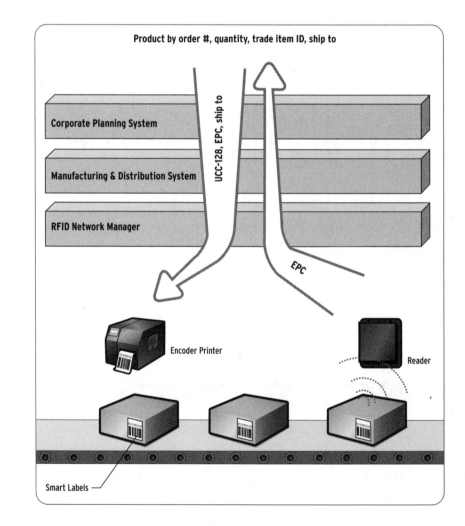

for tracking purposes. You can use a PC to generate EPCs along with other labeling information, and locate it right next to a labeling station. You may find this approach to be the easiest and least expensive for near term compliance driven applications. Label application software companies are now coming out with EPC modules to handle the task. Alternatively, a software module within your supply chain execution system may generate an EPC and embed it in the print job stream. You can also purchase pre-numbered RFID tags. In this situation you would need to set up a system at point of application to create a database that matches GTIN with EPC.

⊕ **SEE PAGE 277**
on using RFID data in the
supply chain.

Trade item data – Serialized trade items in EPC format should link with your master catalog. Options include converting the catalog to a new database schema and storing the new information there, or using a separate software module that does a look up and cross-reference between your master catalog and assigned and available EPC numbers.

Process management – Your distribution system needs to poll readers for tag data, validate the information on packages and pallets, and use that information to trigger next steps in the process.

Exception handling – The process must react to quiet tags, mis-labeled cases, and trigger error recovery routines, alerts and management reports.

ASN – Advanced shipping notice information requires reading of a pallet EPC, correlation of case EPC numbers with the pallet, and EDI transmission of the shipping confirmation.

Management reporting – Records need to be kept to validate the tag encoding, smart labeling and reading process. They become part of the product chain of custody.

Returns – EPC number lookups and handling systems for returns may require modifications to existing systems. Tag readers help automate these systems to reduce special handling labor.

⊙ SEE PAGE 252
on components of an
RFID network.

Middleware – RFID systems create a lot of data. Software components, called middleware, are used to filter, store and forward the data to other systems, thus reducing network traffic. Commercial application integration software can package EPC data with descriptions in PML or some other application-neutral format so it can be made available to supply chain partners using a publish and subscribe protocol.

ENTERPRISE-WIDE SMART LABEL PRINTING MANAGEMENT

Enterprise integration of smart label printers offers another level of capability and control. With print management software, all label printing operations can be monitored from a web browser practically anywhere on the planet. Printers can be made visible to ERP and systems management software applications. Alarm conditions can be set and monitored, diagnostics run, and alerts sent to the appropriate individuals. In addition, remote print management allows centralized configuration control, lockouts of printer configurations, and coordinated updates to firmware. See Figure 4.9.

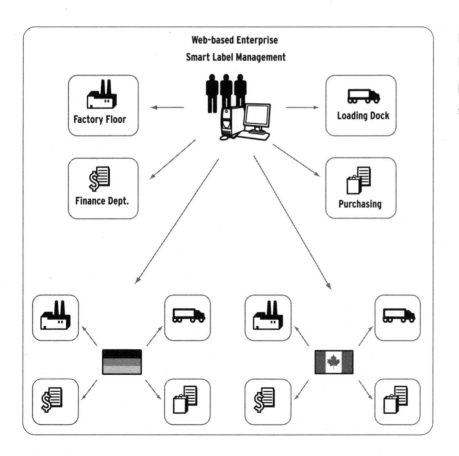

FIGURE 4.9

Enterprise smart label print management system.

Sources and Further Information

RFID for the Supply Chain: Just the Basics, white paper, Printronix, February 2004.

A Basic Introduction to RFID Technology and Its Use in the Supply Chain, white paper, by Steve Lewis, Laran RFID, January 2004.

RFID Primer, white paper, Alien Technology, 2002.

8 Supply Problems when Using Label Printer Applicators, white paper, Fox IV Technologies, available: www.foxiv.com.

General Tag Overview, presentation by Symbol/Matrics, 2005.

CHAPTER 5

Industry Initiatives

WAL-MART 138

Wal-Mart RFID System Requirements 141

Advanced Shipping Notice Integration 142

Replenishment System Data Sharing 143

OTHER RETAIL MANDATES 146

European Retail 147

GLOBAL SUPPLY CHAINS 151

RFID in China 151

Protecting the Global Food Supply 152

Challenges of Ocean Shipping 154

PHARMACEUTICAL INITIATIVES 156

IN THIS CHAPTER:
Retail industry mandates for RFID deployment.

Of the four determinants of a market: availability of technology, formation of standards, reasonable price points, and influential buyers, it's the buyers that have largely shaped the direction of RFID deployment. The early buyers, including Wal-Mart, Target, Tesco and Metro, are investing directly and through their supply chains because they have a real need and an expectation of great rewards in return for the risks they are making. Regulators, however, will soon be making an impact on the market, forcing buyers in greater numbers to leverage RFID's potential use in securing the global supply chain.

In this chapter, and in Chapter 6 on the DoD, we are going to look at what the market leaders are currently doing and their roadmap for the future. It is clear that the momentum is there, and what lies ahead is a period of refinement and an opportunity for many benefits.

WAL-MART

The Wal-Mart RFID journey began early in the decade with a trial involving Procter & Gamble, International Paper, Unilever, Gillette, Johnson & Johnson and Kraft. Scientists from the Auto ID Center and the Uniform Code Council lent support to the three-phase trial.

Phase I, which commenced in late 2001, tested small numbers of products and volumes at the pallet level. Phase II began in early 2002 and tested higher case level volumes and a greater mix. Phase III was designed to test individual items tracking down to the unit level, but was scrubbed in 2003 in favor of other Wal-Mart initiatives.

In June of 2003, Wal-Mart announced to its top 100 suppliers that they would be required to adopt RFID technology by 2005. Wal-Mart met with its top 100 suppliers, along with 37 other suppliers who volunteered, to lay out the specifics of its RFID mandate.

The mandate required suppliers to use RFID tags on their pallets and cases shipped to three Wal-Mart distribution centers in Texas by January 1, 2005. Not every product required RFID tagging during the pilot, only selected items and SKUs. It turned out that 53 suppliers actually shipped RFID labeled cases and pallets as of January 1st, and 94 got there by the end of the month, tagging 65 percent of their items on average. By the beginning of March, 2005, Wal-Mart reported that it had achieved 5.6 million successful RFID reads on 23,753 pallets and 663,912 cases of product.

Wal-Mart is continuing its rollout by involving its next 200 suppliers in shipping RFID labeled cases and pallets to the three DC's and 140 stores that are RFID enabled. During 2005, Wal-Mart plans to roll out RFID to 600 Wal-Mart locations. Figure 5.1 shows the implementation timeline.

(→) SEE FIG. 5.1

Why is Wal-Mart doing this? What do they stand to gain? It turns out that Wal-Mart could save itself billions of dollars, according to a report by A.T. Kearney. Savings will come from improved tracking of supplies, which should lead to a five percent reduction of store inventory requirements. Labor costs for inventory management are projected to drop by 7.5 percent. Wal-Mart expects initial labor efficiencies of ten to twenty percent at its distribution centers. A mere one percent reduction in out-of-stock would translate to $2.5 billion in added annual sales, since Wal-Mart's total annual sales is about $250 billion.

Wal-Mart handles approximately 8 billion cartons of goods annually, and its top 100 suppliers account for 12 percent of that. The RFID initiative therefore, involves tagging and tracking millions of cartons beginning in 2005.

An independent study of Wal-Mart suppliers seems to verify that

FIGURE 5.1

Planned Implementation Timeline.

2004	2005	2006
Pharmaceutical tracking pilot	Top 100+ suppliers tagging for 3 Texas DC's	January – Next top 200 suppliers begin tagging cases and pallets
Refinements to RFID strategy		
May – 21 products from 8 suppliers tagged for Sanger, TX DC and 7 local Supercenters	June – 6 DC's, up to 250 Wal-Mart and SAM'S CLUBs	
	October – up to 13 DC's, 600 stores	

those who took part in the pilot acquired substantial RFID implementation knowledge and practical experience. According to the study, the vast majority of the original 137 suppliers were making efforts to comply during the months leading up to January, 2005. Active resistance to the mandate was uncommon. The financial investment these companies made was much less than analysts had projected. Spending by the 137 companies averaged $500,000.

RFID project teams at the companies were typically drawn from the IT side of the business. The study found that 45 percent of the companies managed the compliance effort by doing it themselves. 24 percent of the companies contracted out the entire effort to a third party compliance vendor. By and large, the least expensive approach was found adequate for this first phase. Very few companies did any significant re-engineering work. Companies favored partnering with smaller SI vendors with deep knowledge of retail supply chain and logistics, rather than the big name IT consulting firms.

LESSONS LEARNED

Wal-Mart RFID System Requirements

The company provided its suppliers with a list of requirements of which this is a summary:

Tags – Rather than wait for Gen 2 tags, Wal-Mart has chosen to accept existing EPC compliant first generation tags. They can be durable, temporary or permanent read-only 64 or 96-bit Class 0,

Class 0+ or Class 1). Wal-Mart expects its suppliers to migrate to Gen 2 tags when performance, availability and price permit. Tags must operate in UHF (866-956 MHz) spectrum.

Tag application criteria – Tags are to be on both pallets and cases, including returnable containers, stretch wrapped bundles, bags and direct-to-store delivery trays. On full case pallet shipments, just the pallet tag will be read, not every case. A successful read of a pallet tag is defined as a minimum three reads at a distance of up to 10 feet from an antenna. Tag read rates of 100 percent are required. For individual cases, a 100 percent read rate is expected when cases are traveling up to 600 feet per minute, with a six-inch separation between cases. Tags must be readable from any of six sides.

⊙ SEE PAGE 214
on reader antenna
placement.

Readers and antennas – Recommended are so-called agile readers that can handle multiple tag classes and frequencies. Readers must be Ethernet-based, have flexible output options, and RF environment awareness, built-in security, and be capable of disabling unused features (web servers, etc.). Dock portals require an antenna on each side of the dock door and an additional antenna above the door. Conveyor curtains require one antenna on each side of conveyor (speed up to 600 feet per minute) for case tagging.

Advanced Shipping Notice Integration

EPC data from case and pallet loads becomes part of the ASN a supplier sends to Wal-Mart ahead of the shipment. ASN involves instantaneous communication, via EDI, once packages and pallets are sealed and labeled by the manufacturer for shipping. The order detail contained in an advanced shipping notice associates case numbers with a specific pallet. It will be matched against the receiving detail (pallet ID code) that is automatically read at the dock door when a tagged shipment arrives (Fig. 5.2). A warehouse management system verifies receipt and directs the next step, such as checking the shipment into inventory or cross-docking for outbound transit.

⊙ SEE PAGE 264
on event management.

⊙ SEE FIG. 5.2

Replenishment System Data Sharing

At each of the 104 Wal-Mart stores and 36 Sam's Clubs, the company has installed RFID readers at the receiving docks at the back of the building, near the trash compactors and between the back room and the retail floor. As RFID tagged cases of product are shipped to the stores, Wal-Mart records their arrival by reading the tag on each case and then reads the tags again before the cases are brought out to the sales floor.

By using sales data from its existing point-of-sales system, Wal-Mart subtracts the number of cases of a particular item that are sold to customers from the number of cases brought out to the sales floor. Based on that information, software monitors which items will soon be depleted from the shelves and automatically

FIGURE 5.2 Advanced Shipping Notice integration with RFID are part of the mandate.

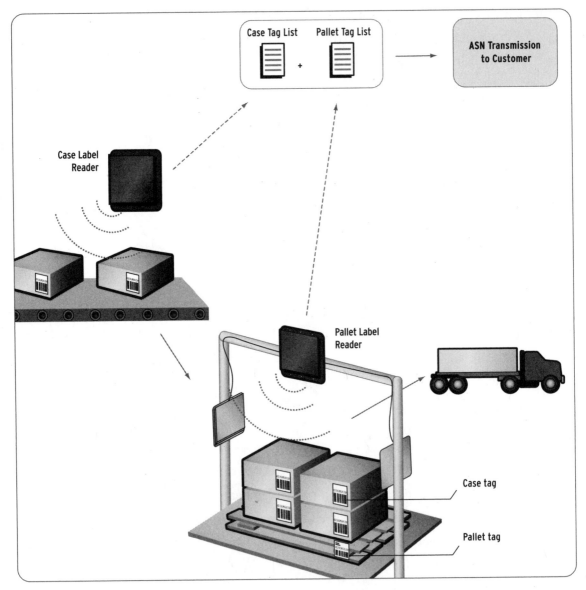

generates a list of items that need to be picked from the back room in order to replenish the store shelves.

Wal-Mart associates use a handheld RFID reader that acts like a kind of Geiger counter, beeping when an associate gets close to the item he or she needs to pick. That reduces the amount of time spent in the back room. The plan is to deploy the handheld devices to associates in 140 stores during the course of the year.

The company is sharing data from all its RFID read points with suppliers through Wal-Mart's Retail Link extranet. When a case is brought out to the sales floor, the system records that it's being put out on the shelves. When the case is read at the trash compactor, the status within the system is changed to "on shelf." Suppliers can get updates on the location of their product within 30 minutes of the system recognizing its movement through the portals.

By integrating data at various read points with its procurement and inventory management systems, Wal-Mart is leveraging the instantaneous nature of RFID and putting it to good use. Similarly, its vendors can use the information being shared in a number of ways:

• Pallet data at receiving docks can provide a confirmation of shipment and time received. By comparing this data with their own logistics data, they can potentially streamline delivery operation, sort out mistakes proactively and reduce chargeback occurrences.

• Visibility into case movement within Wal-Mart provides more up-to-minute information than what is available from a point of sale system, aggregated by shipping quantity and SKU. This can help tailor production plans, replenishment scheduling, and provide early indications of sales promotional performance.

OTHER RETAIL MANDATES

Although Wal-Mart assumed a leadership role by pushing its mandate first, other retailers are close behind and more are expected to make announcements.

Target Corporation, with approximately 1,200 stores in North America, works with its top-tier vendor partners to apply tags to pallets and cases and start shipping to select regional distribution facilities beginning late spring 2005. Target's intent is to accept EPC tags from all vendors as a supplement to the current bar code markings at the carton and pallet level by spring 2007. Target sees RFID as a complement to current bar code and EDI technologies. For the foreseeable future, the current carton marking requirements for shipping containers will remain unchanged.

Albertsons is one of the world's largest food and drug retailers, with 19 DC's and approximately 2,300 retail stores in 31 states across the United States, including Jewel-Osco, Acme, Sav-on

Drugs, Osco Drug, and Super Saver. The company is currently in the testing phase using RFID technology with select partners at the case and pallet level. Albertsons expects its top 100 suppliers to be participating in the RFID program at the case and pallet level by early 2005.

Electronics retailer Best Buy is beginning an implementation across its 830 stores in US and Canada. In its planning and justification study, Best Buy has found RFID to be five times faster than bar code for inventory track and trace work. It anticipates a $400 million payback over a five-year period once it implements item level RFID tagging. Although RFID at the case and pallet level will increase distribution center velocity and save labor costs, the majority of its payback will come from "front of store" operations, including warranty registrations and return handling.

European Retail

Together with Intel Corporation, Europe's three largest retailers, Tesco PLC, Carrefour Group and Metro Group, formed the EPC Product Retail Users Group of Europe. This independent working group complements the efforts of EPCglobal, actively piloting EPC and RFID technologies in their supply chains.

Tesco, the largest retailer in the United Kingdom, and among the most active retailers testing RFID technology, is putting RFID tags on cases and re-usable totes of nonfood items at its distribution centers and tracking them in stores. The rollout to over 900 stores started in 2004. Tesco is using Class 1 866 MHz tags.

UK clothier Marks & Spencer initiated an item-tagging trial that it plans to extend to 53 stores by 2006. The store is tagging men's suits and women's bras with passive UHF tags in order to improve product availability. Responding to consumer concerns about the trial, the retailer promised not to use the tags at the checkout level, or tie tag data to individual customer payment or refund information. The tags are scanned by stock workers using mobile readers at the end of the day to verify stock levels and determine each day's replenishment needs. Replenishment data is transmitted overnight of Marks & Spencer's distribution center. British Telecom is managing the trial together with Intellident and EM Micro.

Metro Group, based in Germany and Europe's third largest retailer, unveiled an RFID "Future Store," showcasing the benefits of the technology for shoppers. Metro Group has over 2000 stores in 28 countries. Its initial rollout involved 20 suppliers. It began a comprehensive pilot along the entire supply chain in 2004, with about 100 suppliers tagging cases and pallets for delivery to ten central warehouses and 250 stores.

In its test store in Rheinberg, Metro is implementing RFID within the store aisles and checkout areas as well as receiving. For inventory management, RFID is used on the shelves to detect low shelf stock, or misplaced items. Customers can scan products for themselves while in the store, and use an intelligent scale to calculate price on weighed items. Display terminals show customers information about products, including personalized ads and suggested recipes (Fig. 5.3). The checkout system quickly totals and displays the price of goods in their shopping cart (Fig. 5.4).

→ SEE FIG. 5.3 and FIG. 5.4

Customers can debit a credit card account or pay the cashier. A de-activator display after checkout allows customers to kill an item tag if they wish (Fig.5.5). Metro is making information on RFID, EPC and the Future Store concept available in the store to help educate their customers. Early surveys at the Future Store show that at least 25 percent of their customers have tested all the new systems themselves. Satisfaction with shopping experience has increased from 34 percent to 52 percent.

→ SEE FIG. 5.5

Carrefour, Europe's largest retailer and second largest in the world, sees global implementation of RFID as assurance that customers can have better product availability and value. Carrefour operates almost 10,400 stores in 30 countries across four formats Hypermarket, Supermarket, Hard Discount and Convenience.

5.3A

5.3B

FIGURE 5.3A

Product information via RFID.

FIGURE 5.3B

Personal ads while shopping.

5.4

5.5

FIGURE 5.4

Checkout portal.

FIGURE 5.5

De-activator.

Photos courtesy of Metro

GLOBAL SUPPLY CHAINS

RFID is part of a global trend that is reshaping the retail supply chain. Wal-Mart appears to be at the forefront of this as well. Wal-Mart has about 6,000 global suppliers, 80 percent of which have operations in China. China is the largest exporter to the US in virtually all consumer goods categories, making Chinese business practices and its regulatory environment pivotal to the new global supply chain.

One of the key characteristics of this new supply chain is the elimination of distribution tiers, a phenomenon known as disintermediation. Wal-Mart contracts directly from its suppliers in China. Its global procurement center is in Shenzhen, a city in southern China having what is becoming the third busiest seaport in the world. In its goal of bringing products to market at the lowest price, Wal-Mart is taking over the supply lines, eliminating smaller and less efficient companies in the middle, and investing in information technology to gain visibility into all aspects of global supply.

RFID in China

Recognizing the rapid change taking place in global supply chains, the Auto-ID Labs has established centers throughout the world as a part of its efforts to provide solutions for the global supply chain. One of its centers is affiliated with Fudan University near Shanghai,

China. It was set up to promote the use of EPC and RFID standards. Similarly, there are labs in Japan, Korea, Switzerland, UK, Austria, and the US.

Clearly, it would benefit retail companies such as Wal-Mart to have RFID tagging done at its source locations in China, but a potential conflict exists if tags designed for use in the US are incompatible with encoding and RF standards permitted in China or elsewhere. Whereas in other parts of the world, private enterprise has set the agenda for RFID adoption and governments took a back seat, in China the government wants to play a more active role. The Chinese government announced its intent to jointly develop an Asian standard for RFID in partnership with Korea and Japan. Currently there are three Chinese government ministries, one having to do with allocating and licensing the radio spectrum, another with numbering systems, and a third on trade, involved with setting RFID standards. Although progress is being made, it is not clear whether the Gen 2 standard will be prevail, even with pending ISO adoption.

Protecting the Global Food Supply

Global supply chains for food and beverages have a special set of challenges because of regulations aimed at protecting the food supply. The US Centers for Disease Control estimates that 76 million illnesses and 5,000 deaths occur annually due to contaminated food, translating into annual health care costs of $6.5 billion. Japan's Food Safety Basic Law, and certain provisions of the US Bioterrorism Act, put in place following the outbreak of "mad cow" disease, requires site registration, and product labeling, tracking and tracing provisions that are affecting thousands of importers.

To help protect its meat trade with Japan, all cattle in Australia are having RFID tags embedded in them to track the meat from calf to butcher. Similar track and trace requirements are in place in Europe and Canada. China Daily, a daily news portal, reported in a May 2004 article that the FDA's Bioterrorism Act will affect China's businesses significantly. More than 8,000 Chinese businesses already have registered with the FDA as required by the Act.

The FDA published its final ruling on the "Establishment and Maintenance of Records," Section 306 of the Bioterrorism Act in December, 2004. The intent of the ruling is to implement an ability to trace forward and quickly remove adulterated food that poses a significant health threat. Although RFID is not specifically mandated, it has obvious benefits.

→ SEE PAGE 253 on regulatory and operational requirements.

The rules call for prior notice of import of foods, within four hours or "wheels up" in the case of air shipment, and 24 hours prior to landing for ocean shipment. The records availability requirement has been revised to "as soon as possible" not to exceed 24 hours from the time of receipt of an official request. The FDA expects "as soon as possible" to ensure rapid response to any situation. An FDA risk assessment study done in 2003 states that "if an unintentional contamination of one food, such as clams, can affect 300,000 people, a concerted, deliberate attack on food could be devastating, especially if a more dangerous chemical, biological or radio-nuclear agent was used."

Challenges of Ocean Shipping

Oceanic shipping presents a number of issues for global supply chains. According to a study by A.T. Kearney, twenty thousand containers enter US ports every day, and more than 95 percent go uninspected. The US Bureau of Customs and Border Protection is fast at work tightening the borders, and its mandates will dramatically affect the global supply chain.

For food, the US Bureau of Customs and Border Protection requires a stringent and detailed advanced shipment notification and electronic presentation of cargo information. The manufacturer, processor or packer must keep a record of lot or code number or other identifier if the information exists. The FDA wants the record to include all information "reasonably available" on the specific source of each ingredient used to make the finished product.

LESSONS LEARNED

RFID may offer a way to meet the track and trace mandate, although it also presents another set of problems such as when stacks of tagged product rub against each other over many hours, causing

ESD discharge and abrasion that can damage tags.

Customs-Trade Partnership Against Terrorism (C-TPAT) guide-lines, the latest version of which was published in March, 2005, affects all US importers. C-TPAT establishes mandatory security standards, specifying actions importers must take to ensure the security of warehouses, container storage and loading, the hiring of personnel, information technology, and supplier security practices. Phased in over a 120 day period from March 25, 2005, are the procedural security rules regarding document processing, manifest procedures, shipping and receiving and cargo discrepancies.

The number one concern of top 100 importers and exporters is that of assuring container security according to the A.T. Kearney study. The study makes a clear link between RFID and the strategy of importers to meet the C-TPAT guidelines. RFID can help importers demonstrate container security and safeguard their rep-utations in a time of crisis. Among other validation measures, smart box technology, defined in the study as active RFID tags used in container seals, and passive tags on cases and pallets, will earn C-TPAT participants "green lane" status, speeding security clearances at ports. Although not mandated, it is clear that RFID can be a tool to secure and streamline the global supply chain in the face of increased government scrutiny and potential disruption from terrorist activity.

PHARMACEUTICAL INITIATIVES

The US Food and Drug Administration is actively promoting the use of RFID to improve the safety and security of the drug supply. The FDA goal is to have RFID adopted by the drug supply chain by 2007. This follows a 2004 pilot by Wal-Mart that was sanctioned by the FDA. The potential application of RFID falls within a set of proposals and recommendations involving electronic pedigree to guarantee the authenticity of drugs. RFID is viewed as having potential, because tags aren't easily tampered with, and the EPC code structure enables the assignment of unique identifiers throughout the process.

Prescription drugs have a rather complex supply chain, involving contract manufacturing, wholesaler intermediaries, and repackaging as drugs find their way to a retail counter. Counterfeiters have found multiple entry points into that chain to substitute drugs. The high retail value of certain breakthrough drugs has created high risk categories, susceptible to counterfeiting. Counterfeiting costs the industry almost $30 billion within the $327 billion global drug market. Drug companies faced almost 1,300 product recalls in 2001, and more than $2 billion yearly in product returns caused by overstocked or outdated drugs.

→ SEE PAGE 261

on application level events (ALE).

Electronic tracking technology, such as RFID, along with "chain of custody" business practices, could make it much more difficult for illegitimate and rogue operators to develop entry points within the distribution supply system. Participants in the Wal-Mart pilot have found that RFID has demonstrated cost effectiveness by improving inventory control, expediting delivery shipments and reducing product waste and diversion.

The recommended approach across the supply chain is called "one forward, one back." This approach is based on the concept that at each point in the supply chain, a certification is made that the drugs were received from a valid source (one back), and will be shipped to a valid source (one forward). RFID implementation is likely to be phased in, starting with the serialization of bulk quantities.

The FDA issued guidelines in 2004 for RFID feasibility studies and pilot programs, making it easier for drug companies to begin testing. Drug makers no longer need to request special authorization. The guidelines limit testing to prescription or over-the-counter drugs, and state that RFID tagging is not a substitute for current FDA package labeling requirements and control systems.

What's being tested currently are bottle caps and packet labels with high-frequency passive (13.56 MHz) tags, and passive UHF tags on bulk quantities. The current debate is whether HF tags, with their smaller size and shorter read range, are more appropriate for item-level drug tagging than UHF tags. 13.56 MHz tags are read at close distance using near-field EM. The longer HF wavelength is less susceptible to absorption by liquids. 13.56 MHz chip and antenna configurations can conform to a 9mm or smaller diameter, suitable for injection molding into bottle caps. What UHF may have in its favor is lower cost if mass adoption of Gen 2 technology moves forward.

A number of drug companies have started trials. Pfizer announced its plans to place RFID tags on all bottles of Viagra intended for sale in the US in 2005. GlaxoSmithKline announced that it intends to begin using RFID tags in 2005 on at least one product deemed susceptible to counterfeiting. Purdue Pharma announced that it is placing RFID tags on bottles of OxyContin to make it easier to authenticate as well as track and trace the pain medication. Johnson & Johnson has also been active in trialing RFID as a part of the Wal-Mart pilot.

EPCglobal has an operational action group in the healthcare and life sciences industry, supporting the pilots of passive HF and UHF tags. The Healthcare Distribution Management Association recently took the position that manufacturers and wholesalers should use EPC tags at the case level, with a goal for deployment of December 2005. In addition, HDMA is suggesting that drug packagers and manufacturers should set a goal of deploying EPC tags at the item level by 2007.

Sources and Further Information

"Wal-Mart's Roadmap to RFID," by Dave Kelly, *RIS News,* January 2004.

Meeting the Retail RFID Mandate, white paper, A.T. Kearney, November 2003.

Understanding the Wal-Mart Initiative, white paper, Matrics Systems Corporation, 2004.

"Wal-Mart Begins RFID Process Changes," by Mark Roberti, *RFID Journal,* February 2005.

Wal-Mart's RFID Deployment – How is it Going? white paper, Incucomm, Inc. 2005.

"EPC/RFID in Action at Best Buy," presentation by Paul Freeman, Best Buy, Minnesota Technology Inc. Awareness Forum, March, 2005.

"Taking Food Safety to the Next Level," *Automation Today* magazine, Rockwell Automation, November, 2004.

"China's Changed Everything For Consumer Goods Makers," by Lara L. Sowinski, *World Trade,* January 2005.

Smart Boxes, RFID Can Improve Efficiency, Visibility and Security in the Global Supply Chain, white paper, A.T. Kearney Inc., 2005, available: www.atkearney.com.

"Background to Marks & Spencer's Business Trial of RFID in its Clothing Supply Chain," news release, February 2005, available: www.marksandspencer.com.

"Metro Group To Introduce RFID Across The Company," news release, December 2004, available: www.metrogroup.de.

Radiofrequency Identification Feasibility Studies and Pilot Programs for Drugs, Sec. 400-210, November 2004, available: www.fda.gov.

"FDA wants RFID along on drug trips," *DC Velocity,* January 2005.

"Beyond the Bioterrorism Act," by Jon Blanchard, *Food Engineering,* May, 2005.

"FDA Announces New Initiative to Protect the U.S. Drug Supply Through the Use of Radiofrequency Identification Technology," news release, November 2004, available: www.fda.gov.

Radiofrequency Identification Feasibility Studies and Pilot Programs for Drugs, Compliance Policy Guides, U.S Food and Drug Administration, November 2004.

Item-Level Visibility in the Pharmaceutical Supply Chain, A Comparison of HF and UHF RFID Technologies, white paper by Philips Semiconductors, TAGSYS and Texas Instruments Inc., July 2004.

CHAPTER 6

Department of Defense Initiative

SENSE AND RESPOND LOGISTICS   163

RFID's Role in the DLA Transformation   165

GUIDELINES FOR IMPLEMENTATION   169

Updated Mil Standard for Package Labeling   171

Advanced Ship Notice   173

Gen 2 Transition   174

Data Constructs   174

Supplier Funding for Implementation   176

IN THIS CHAPTER:
Defense Department mandates for RFID deployment.

If ever there was a supply chain in need of what RFID promises to bring, it's that of the Defense Logistics Agency. The scope of the agency is mind boggling. If the DLA was an enterprise, it would be the second largest distributor in the world. It has $28 billion in sales and services, over 22,000 civilian and military personnel, and locations in 48 states and 28 countries. It processes 8,200 requisitions a day, and has over 45,000 suppliers and 5.2 million items in supply.

It is not the metrics that best describes the defense logistics challenge, however; it's the mission. The DLA needs to support rapid US troop deployment and sustained warfare anywhere in the world. Unlike a retail enterprise, the DLA doesn't get to select its locations, either. It has to be able to extend the supply chain, on a moment's notice, to places that often lack suitable infrastructure, and do it under fire. And it supports a very demanding organization. As Secretary of Defense Donald Rumsfeld said in March, 2003, "To win the global war on terror, the armed forces simply have to be more flexible, more agile, so that our forces can respond more quickly."

SENSE AND RESPOND LOGISTICS

The current ongoing mission in Iraq illustrates the challenge of the new "sense & respond" logistics capability being demanded by the Department of Defense. Every day there are thousands of items needing repair coming out of Iraq, and thousands of repaired items going back. Planners need to know where the parts are, what transportation systems are moving them along and what competing priorities there are to getting them fixed. To do that they need complete visibility in the supply chain, including new and repaired parts supply systems, procurement systems and delivery timing. In addition, they need some way of integrating other types of planning and event prediction information, such as failure rates for new and repaired parts. Because practically everything needs to be airlifted to Central Asia, the military air transport system represents a severe bottleneck that has to be carefully managed.

The DoD's new logistics strategy is fairly recent and far reaching. About twenty years ago the DLA maintained warehouses full of inventory to support the military. Then it began adopting Japanese just-in-time business practices in partnership with its suppliers, relying more on daily deliveries. But in parallel with this effort to reduce buffer stock, weapons systems became more complex. The US Air Force, for example, with its advanced fighter aircraft, was becoming overwhelmed with having $33 billion of parts inventory in the late '90s.

During the first Gulf war in 1991, nearly one-third of the containers shipped to the Middle East were lost or unaccounted for when needed. Those that did arrive had to be manually opened to see what they contained. Battlefield commanders began to distrust the supply chain and made redundant requisitions in a practice they called "just-in-case" logistics. The result was "iron mountains" of containers on docks and in the desert, providing an expensive lesson to war planners and the impetus to a new logistics paradigm.

Today, the DoD practices Performance Based Logistics (PBL), in which they contract for a need rather than specific materials. According to James D. Hall, a senior official in the DLA, "The government will not own spare parts, but pay for secure levels of availability." While the size of DLA sales and services has grown by 70 percent over the last three years, its personnel levels have shrunk, to the lowest level since 1963.

PBL is a transformation strategy aimed at redefining the link to the customer. In order to build flexibility within a reduced cost structure, PBL seeks to push more decisions down the chain of command, to improve the response to battlefield command. Information technology is a key enabler to the transformation, which is why RFID is getting a lot of attention.

RFID's Role in the DLA Transformation

RFID is seen as a key enabler of an integrated DoD supply chain. The goal is to increase troop confidence in the reliability of supply. It is aimed at improving the visibility of information and assets, at improving the process of shipping, receiving and inventory management, and at reducing cycle time.

The Department of Defense has been using RFID over the past 10 years, mainly active tag systems for the identification of large containers. It is now the largest operational tagging system in the world. Active tags were on 45,000 pallets of material going to Central Asia to support the Iraq war effort.

Active tags on pallets are read and tracked by the InTransit Visibility (ITV) network as they move through "choke points" in the supply chain. See Figure 6.1. ITV stretches across more than 1,600 locations in more than 45 countries. Used in both Afghanistan and Iraq, ITV has reduced overall losses to less than eight percent. During Operation Iraqi Freedom, the military deployed 90 percent fewer containers compared to Operation Desert Storm. The chief of the DoD Logistics Automatic Identification Technologies office estimated that the military saved $300 million in Iraq through the improved visibility gained by RFID applications.

 SEE FIG. 6.1

DEPARTMENT OF DEFENSE INITIATIVE:
Sense and Respond Logistics

The DoD began trialing UHF passive tags on cases and pallets in 2003, with selected suppliers under a voluntary program. An initial implementation was completed at two distribution depots in Susquehanna, PA and San Joaquin, CA in January 2005. The DoD set up portals to read Class 0 and Class 1 tags, and integrated the information into its DLA warehousing system.

Moving forward, the DoD will begin requiring suppliers to use passive RFID tags on shipments as new contracts are issued to them. In order to do this the DoD needs to create legally binding Defense Federal Acquisition Regulation Supplements (DFARS) in cooperation with the Office of Management and Budget. This is turning out to be a lengthy process, but contracting is expected during 2005, as it plans to implement passive RFID at 16 additional sites.

FIGURE 6.1

DoD supply chain uses the InTransit Visibility (ITV) system to track items from factory to foxhole.

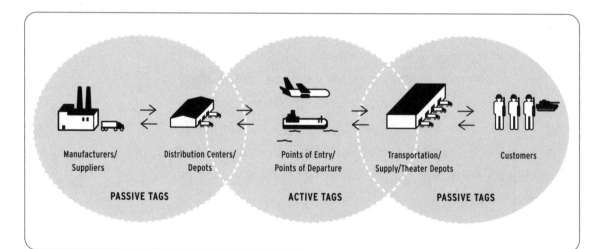

Manufacturers/ Suppliers	Distribution Centers/ Depots	Points of Entry/ Points of Departure	Transportation/ Supply/Theater Depots	Customers
PASSIVE TAGS		**ACTIVE TAGS**	**PASSIVE TAGS**	

We'd like to hear from you

As thick as this book is, it may not have the answers to all of your questions. It also may not include everything you'd like to see. So your input will help make the next edition of RFID Labeling even better. Thank you.

My comments, questions, corrections, etc.:

The most useful part of this book is:

The least useful part of this book is:

☐ I have an application I'd like to discuss.

☐ I'd like to have a product demonstration.

☐ I'd like to attend a seminar on Printronix products.

Name

Title

Company

Address

City   State

Zip   Phone (   )

Email   **PRINTRONIX**

BUSINESS REPLY MAIL
FIRST-CLASS MAIL PERMIT NO. 5746 IRVINE CA

POSTAGE WILL BE PAID BY ADDRESSEE

Marketing Department
PRINTRONIX
PO BOX 19559
IRVINE, CA 92623-9559

Although the DoD intends to adopt EPC standards and leverage the retail industry initiatives of Wal-Mart and others, there are differences in how the DoD is managing its rollout. The DFARS contracts, for example, allow suppliers to factor the costs of tags into their pricing. The DoD will not, however, underwrite the investments a company such as GE or Boeing will make to implement RFID internally to gain efficiencies. Unlike Wal-Mart, the DoD is selecting suppliers initially based on the value of goods, rather than the amount of business being transacted.

→ SEE PAGE 345
case study on a military supplier.

The DoD has 45,000 suppliers, of which 500 are supplying most goods. The DoD supply chain is much more complex than that of a retail company. It involves multiple autonomous branches of government and everything from toiletries to tanks. Practically all of Wal-Mart's top 100 suppliers also supply to the DoD. There is a great deal of overlap with the rest.

The DoD differs from retail in its operating environment as well, and represents a challenge to RFID technology. Much of what the military consumes is made of metal – bullets, guns, armor, etc., which interferes with radio signals. Radio signals can trigger certain types of ordinance, necessitating separate tag certification guidelines. Tags will need to be able to withstand very harsh environments including the high temperatures of a desert, the grit of a sand storm, and the humidity of a jungle. Lastly, a volunteer army under the strain of combat will increase the failure rate of any system that depends on sensitive electronic components, their precise orientation and predictable movement in order to function reliably.

LESSONS LEARNED

A 2004 field study seems to indicate that the reliability of passive UHF is in question. The study tested the read rates of passive UHF tagged cases and pallets as they were transported through various airlift site and forward staging areas. During the study, 93 percent of the shipping label tags were read at least one time through the network of readers. Read rates at individual portals were much lower, however. 76 percent of the label tags were read at Dover Air Force Base, prior to consolidation for shipment. Once airlifted to Ramstein AFB in Germany, the palletized read rate was 68 percent. After deconsolidation, only 38 percent of the tags were read.

One reason for the study was to determine RFID capabilities within the existing environment. No attempt was made to optimize read rates by changing existing logistics processes or tag placements. The study concluded that without changes to the current process, passive RFID will not achieve a 100 percent read rate. Given a system with less than 100 percent read rate, the study concludes that RFID will augment the in-transit verification process but still not replace other sources of data.

Clearly, if RFID cannot reliably replace existing systems, it will not fully deliver the benefits of speedier information flow, nor labor and infrastructure savings. The DoD nonetheless continues to maintain its schedule of implementation. It is fully funding its passive RFID initiative, and has the full commitment of senior leadership. The DoD expects to save over $464 million over the next six years through RFID. The DoD has launched an outreach program to help its suppliers meet its RFID requirements, including making presentations at industry conferences and providing education programs. In a survey of 400 of its suppliers in November, 2004, the DoD reported that 78 percent of respondents had plans to integrate RFID into their business processes and 82 percent planned to comply with EPC standards.

GUIDELINES FOR IMPLEMENTATION

The DoD released its final policy on passive RFID in July, 2004. The effective date depends on clauses written into each supplier contract upon renewal. DFARS are usually renewed every two years, so it is expected that by 2007 all suppliers will have some type of RFID contract clause. January 1st, 2005 was the starting date for voluntary compliance. See Figure 6.2.

⊙ SEE FIG. 6.2

DoD guidelines leverage EPCglobal standards, so as to share in the benefits of other initiatives including that of Wal-Mart. In parallel to the Auto ID Center work, the DoD recognized the need for harmonizing code data, since it affects 1,500 logistics systems. The guidelines stipulate the use of a unique identification (UID) number at the item-level to track assets of value. The intent is for UID and EPC to be compatible and the UID system implementation program in full swing by 2007.

Commencing January 1, 2005 – For suppliers with new contracts, case and pallet tagging is expected for rations, clothing, individual equipment (helmets, boots, etc.) weapon system repair parts and components and personal items (Wal-Mart items) being shipped to the Susquehanna, PA and San Joaquin, CA depots.

FIGURE 6.2

DoD implementation

timeline.

2005	2006	2007
Voluntary participation for tagged shipment to Susquehanna, PA and San Joaquin, CA	Mandated participation for tagged shipment to 34 depots	Everything shipped to any DoD location must be identified with an RFID tag
Class I subclass packaged operation rations	Class I subsistence and comfort items	
Class II clothing, individual equipment and tools	Class III packaged petroleum, lubricants, oils, preservatives, chemicals and additives	
Class VI personal demand items	Class IV construction and barrier equipment	
Class IX weapons systems repair parts and components	Class V ammunition of all types	
	Class VII major end items	
	Class VIII pharmaceuticals and medical materials	

Commencing January 1, 2006 – Case and pallet tagging is required for subsistence and comfort items, packaged petroleum, oils, lubricants, preservatives, chemicals, additives, construction and barrier material, ammunition, pharmaceuticals, medical materials and major end items going to 32 additional distribution depots in the US.

Commencing January 1, 2007 – RFID tagging on cases, pallets and UID items will be required for all DoD suppliers who have new contracts, for all classes of supply to any DoD location.

Excluded from the RFID tagging requirements are bulk commodities carried in rail tank cars, tanker trucks or pipelines. Examples include sand, gravel, water, liquid chemicals, petroleum, ready-mix concrete or similar construction materials, combustibles such as firewood, and agricultural products including seeds, grain and animal feeds.

Updated Mil Standard for Package Labeling

Mil-Std-129P, the military standard for package labeling, was updated in 2004 with "Change 3" specifying RFID tagging requirements. The tagging requirements by RFID layer are shown in Table 6.1. The new standard contains definitions, tag data standards, tag placement guidelines and performance requirements. The standard includes the following:

⊖ SEE TABLE 6.1

DEPARTMENT OF DEFENSE INITIATIVE:
Guidelines for Implementation

- **Approved frequency range** – 860-960 MHz

- **Readability at portals** – For palletized unit loads, exterior containers within the palletized unit load, and UID item unit packs with passive tags that are passing through a portal, the readability needs to be 100 percent at 3 meters (3.3 yards), while traveling at 10 miles per hour.

- **Readability at conveyors** – For individual cases or UID packs with passive tags moving on a conveyor, readability needs to be 100 percent at 1 meter (1.1 yard) while traveling at 600 feet per minute.

- **Labels (Smart Labels)** – Tags can be integrated with the military or commercial shipping label, or the tags can be applied separately so long as they do not interfere with use of the existing label. Both linear and two-dimension bar codes must appear on

TABLE 6.1

Tagging requirements by RFID Layer.

RFID Layer	Description	Tag Type	Class Tag	Frequency	Read Range	Starting January 1, 2005
0	Item	Passive	0, 1 or higher	UHF	3 m required	Not yet
1	Item package	Passive	0, 1 or higher	UHF	3 m	Required on UID & specified items
2	Transport Unit, Case	Passive	0, 1 or higher	UHF	3 m	Required
3	Unit Load, pallet	Passive	0, 1 or higher	UHF	3 m	Required

military shipping labels, and Code 39 bar codes will continue to be required on interior packages and on shipping containers. The printed part of a smart label must include both human readable and machine readable (bar code) information as a backup to the RFID tag.

- **Tag placement** – Whether or not it is integrated with a shipping label, the tag is supposed to be located in such a way that it can be associated with relevant bar coded information.

- **Tagging of munitions** – Ordinance should not be tagged without being certified for electromagnetic effects on the environment (E3), hazards of electromagnetic radiation to ordinance (HERO), hazards of electromagnetic radiation to fuel (HERF) and hazards of electromagnetic radiation to personnel (HERP).

Advanced Ship Notice

A receiving report called an Advanced Ship Notice is required with all tagged shipments going to the DoD. The intent of the ASN transaction requirement is to pre-populate an association of RFID tag data with the appropriate shipment and material data. The transaction facilitates the identification of material solely on reading the RFID tag.

A key benefit for suppliers is that the payment cycle is initiated upon receipt of an ASN, rather than upon receipt of shipment. This accelerates the payment by several days or more. In general, the ASN contains order information, product description, physical characteristics, type of packaging, marking, carrier information and the configuration of goods within the transportation equipment.

For the system to work, the ASN must be received at the shipping destination prior to the receipt of goods. The ASN can be transmitted by EDI, through the web-based Wide Area Work Flow electronic commerce system (https://wawf.eb.mil/), or through Secure File Transfer Protocol (SFTP). The SFTP option allows the supplier to define the format of the ASN documents.

Gen 2 Transition

The DoD will mandate the Gen 2 tag standard when tags become available and meet compliance and interoperability testing. Until that time, tests, pilots and initial implementations should proceed using the currently available EPC Class 0 and 1 tags. The DoD has published sunset (expiration) dates for the Class 0 and 1 tags (64 and 96 bit) that generally extend their usage into 2007.

Data Constructs

The DoD recognizes two number systems, EPC and their own system.

Since EPCglobal requires you to join the association, the DoD is allowing its suppliers the option of not using the EPC numbering system. Many DoD suppliers do not have commercial sales, and therefore no interest in joining EPCglobal. Data constructs particular to the DoD include:

CAGE – The Commercial and Government Entity code is the government issued supplier identification number. This 5-character code is used instead of an EPC construct for a SGTIN, GRAI, GIAI or SSCC.

Figure 6.3 is an example of a 96-bit tag data construct. Note that the filter field is used to specify whether the tag is on a pallet, case or UID item.

HEADER	FILTER	GOVERNMENT MANAGED IDENTIFIER (CAGE)	SERIAL NUMBER
8 bits	4 bits	48 bits	36 bits

Header – specifies that the tag data is encoded as a DoD 96-bit tag construct, use binary number 1100 1111

Filter – Identifies a pallet, case, or UID item associated with tag, represented in binary number format using the following values:
• 0000 = pallet
• 0001 = case
• 0010 = UID item
• 0011 = reserved for future use

FIGURE 6.3

DoD 96-bit tag data construct.

Supplier Funding for Implementation

The DoD recognizes the cost burden placed on its suppliers, and expects to approve contract changes to absorb some costs. According to Attachment 2 of the Policy letter dated February 20, 2004, "Working Capital Fund activities providing this support will use the most current DoD guidance in determining whether operating cost authority (OA) or capital investment program (CIP) authority will be used to procure the required RFID equipment."

Sources and Further Information

Memorandum, Radio Frequency Identification (RFID) Policy, by Michael Wynne, Acting Deputy Under Secretary of Defense, Acquisition, Technology and Logistics, DoD, October 2, 2003.

Memorandum, Radio Frequency Identification (RFID) Policy – UPDATE, by Michael Wynne, Acting Deputy Under Secretary of Defense, Acquisition, Technology and Logistics, DoD, February 20, 2004.

Memorandum, Policy for Unique Identification (UID) of Tangible Items – New Equipment, Major Modifications, and Reprocurements of Equipment and Spares – USD (AT&L), July 29, 2003, available: www.acq.osd.mil/dpap/UID/index.htm.

DoD Radio Frequency Identification (RFID) Survey, Sept-November 2004, Office of the Under Secretary of Defense for Acquisition Technology and Logistics, available: www.rfid.org.

Statistical Analysis Report for USAF Passive Radio Frequency Identification Military Shipping Label (RFMSL) Initial Capability, version 1.0.1, prepared by Northrup Grumman, February 25, 2005.

Briefings, 2005 DoD RFID Summit for Industry, February 9-10, 2005, available: www.rfid.org.

RFID Monthly, An Overview of RFID Industry Developments, February, 2005, Robert W. Baird & Co.

United States Department of Defense Suppliers' Passive RFID Information Guide, Version 7.0, available: www.rfid.org.

MIL-STD-129P w/CHANGE 3 29 October 2004 SUPERSEDING MIL-STD-129P w/CHANGE 2 10 February 2004, available: http://assist.daps.dla.mil/online/start/.

"I Want You to Tag Your Shipments," by Bob Violino, *RFID Journal,* January/February, 2005.

"What the Military is Teaching Us About Supply Chains," by Neil Shister, *World Trade,* December, 2004.

CHAPTER 7

Smart Start to RFID – 4 Friendly Phases

PHASE 1: GETTING STARTED 180

Establish the Team 181

Feasibility Study 182

Test Lab 186

Producing Labels for Testing 188

PHASE 2: TEST AND VALIDATION 192

System Integration 192

Validate Vendor Choices 193

Point-to-point Testing 194

PHASE 3: PILOT IMPLEMENTATION 194

PHASE 4: IMPLEMENTATION 198

VENDOR CHECKLIST 202

Certification Programs 203

Even if it weren't a requirement for doing business with Wal-Mart and DoD in 2005, RFID is poised to transform supply chain operations over the next decade. This chapter covers the basics of how to establish an RFID program in your organization. Like any new project, its success depends on leadership, proper planning, engaging all the stakeholders, demonstrating early success, and moving forward with an intelligent and detailed plan.

PHASE 1: GETTING STARTED

Along with the promise of RFID come the challenges in implementing it. Navigating your way through a complicated new system that requires its own hardware and software is a daunting mission. Combined with the complexities of evolving standards, converting today's bar codes to tomorrow's electronic product codes (EPC), and the prospect of how all of this changes the way your company functions – it is easy to understand why you might take a long look before you make the leap to RFID. But there are benefits for those who embrace the technology now, and it is possible to start smoothly and slowly, one step at a time through an achievable four-phase plan.

Three activities need to happen, more or less in parallel, during your getting started phase:

Team – You'll need to form a core group, aligned with an executive sponsor and representatives of all affected functional areas in the organization.

Feasibility study – Develop a plan that identifies and addresses the business goals, priorities, dependencies, costs and success measures.

Test lab – Begin technology testing to accelerate your understanding of what's practical and possible.

Establish the Team

Your team should have a strong leader. The internal team should not be expected to do it all. Rather, it should quickly gather and identify the resources needed to get started. An initial roadmap that identifies critical milestones, especially those related to compliance mandates, will help the team identify gaps in resources and anticipate needs well in advance. The team will eventually have to draw upon people from engineering, distribution, IT, procurement and other areas, both inside and outside the company, to accomplish the work.

A recent study of tier 1 Wal-Mart suppliers during 2004 found that internal teams could accomplish a lot on their own to meet a compliance mandate. The team's executive sponsor can support the team by helping to find resources and facilitate communications

LESSONS LEARNED

and cooperation as the project involves other functional areas in the organization. The internal team should get educated on the technology and the process. Initial training can be informal. Eventually, the team needs to sponsor training sessions for packaging people, shipping people and others who will be working day to day with the technology.

It's essential that external members of your team are leaders in the field. Vendors who belong to EPCglobal are a good choice. They work hand-in-hand with the leading retailers, suppliers and DoD. As a participant in EPCglobal, a vendor can better understand your business requirements for effective implementation, can influence developments and progress by integrating technical, cost and performance needs into standards, and will keep your organization's program up-to-date.

Feasibility Study

Elements of a feasibility study include an impact analysis, business case and initial implementation plan for each area of the business. You may want to look at simulation software to model the distribution center and packaging line processes, and be able to work through various scenarios. See Figure 7.1. Scenario one may be a diverted packaging line, or compliance "slap & ship" operation. Scenario two could model an in-line case and pallet labeling operation within the manufacturing line that could handle a higher percentage of SKUs.

→ SEE PAGE 227
on smart labeling
approaches.

Scenario three could model your up-stream supply chain that includes work in process and vendor compliance labeling. Each scenario entails separate decisions regarding infrastructure investment, techniques and approaches. Regardless of the simulation, modeling or analysis tools used, the impact analysis needs to estimate the annual labor cost, labor force count, cycle time and throughput for the process.

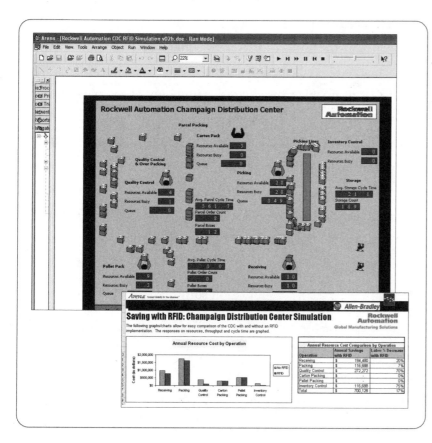

FIGURE 7.1

Simulation software used for impact analysis.

Other elements of a impact analysis may be beyond the capability of the core RFID team, and can be handled by an outside expert. For example, a passive and active RF site survey will help the team understand and mitigate sources of interference. You can avoid a lot of frustration by knowing whether certain elements of your work environment, such as a nearby power transmission substation or cell phone tower, will work against your project plans.

LESSONS LEARNED

The team needs to build a business case for each application, regardless of whether the project is driven by a mandate. A supplier mandate may dictate SKUs and labeling standards near term, but you may find other SKUs justifiable from an ROI standpoint. High value products that experience lots of shrinkage may be the best candidates for an initial pilot, because track and trace through their whole lifecycle could be justified. According to AMR Research, a $15 product cost seems to be the magic number where tagging can be justified. Products and packaging have various sensitivities to RF interference. It would make sense to initially choose an "RF friendly" product, one where tag location is less critical, and the RF signal can penetrate materials and tags can be read in a variety of case orientations. See Figure 7.2.

→ SEE PAGE 246
on a pragmatic approach.

The business case for each product will also help define the "tipping point" when investments in automation and upstream use of RFID are justified. See Figure 7.3. Your company's financial consultant may be able to identify opportunities for tax savings associated with

→ SEE FIG. 7.3

research credits, or grants to offset training costs associated with introducing new technology into your organization.

The implementation plan is the roadmap for the resource expenditures and process improvements over time. The plan should detail the competencies needed, and what can be done in-house versus with outside assistance. Lessons learned by early participants in the DoD program for RFID include that you can outsource certain portions of a project but should not try to outsource process ownership. The internal RFID team needs to manage the process and changes taking place within an organization. Another lesson learned is the need for early engagement of the procurement and IT groups. They need early participation in the implementation plan,

LESSONS LEARNED

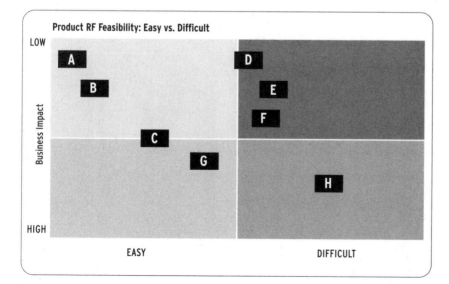

FIGURE 7.2

Prioritizing RFID applications by business impact and ease of implementation.

since they are standards driven organizations with a lot to gain, and with process and infrastructure investments to maintain. Another piece of advice from early adopters is to avoid over-investing when first starting out. You will learn quickest if you minimize causes of variability in a process by keeping it simple. One early adopter suggests that you assume that most equipment will be obsolete in a year and will have to be replaced.

Test Lab

You will want to set up a development environment for small-scale, controlled testing. You may want to create your own test center

FIGURE 7.3

A business case needs to formulate when volumes and labor cost savings tip in favor of production line investments.

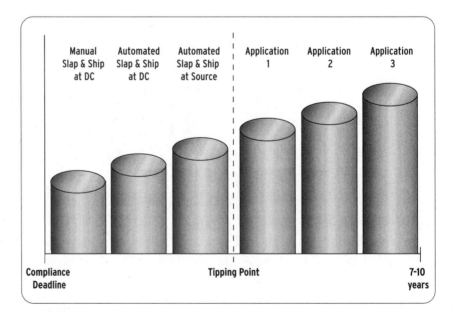

near the packaging shipping area, or use the services of a system integrator who might have their own lab. Equipment that needs to be evaluated and tested includes:

(→) SEE PAGE 209
on lab testing.

- Tags and labels (individual and roll stock)
- Printer/encoders and applicators
- Readers and antenna arrays
- Network infrastructure
- Middleware for data management and device administration
- Conveyor system

Spend sufficient time to research vendors for your lab equipment; as you progress through the implementation stages, you will not want to be changing technology vendors as this will introduce new variables. Choose equipment manufactured by industry leaders who have demonstrated success in broad deployments of their RFID products and that have concrete programs to protect your investment as technology evolves.

One valuable lesson coming from Kimberly-Clark, a tier 1 Wal-Mart supplier, came out of their test lab. K-C created a 'dirty' lab, locating it near their actual production lines. The lab had the same amount of microscopic dust that was present in the production area, and numerous sources of RF interference, including forty wireless terminals nearby. K-C knew that RF tagging of their baby wipes product would be challenging to begin with, because of the

LESSONS LEARNED

moisture content of the wipes and the ESD from the plastic bagging used in the packaging. By intentionally replicating a dirty environment, K-C felt it could identify and isolate associated problems more readily. As it turns out, a number of locations on packages were evaluated before K-C arrived at the right one for roll-out.

Producing Labels for Testing

→ SEE PAGE 352

for Printronix Smart Label Developer's Kit.

A logical starting point is to convert your existing bar code labeling process to one that produces RFID smart labels. A Printronix Smart Label Developer's Kit, for example, can produce smart labels right out of the box. The kit includes software migration tools that provide a seamless transition to encoding and printing smart labels without incurring high reprogramming costs, waiting for official EPC numbers, or changing anything within your front end or back end applications.

The kit includes a suite of applications that convert standard UPC and Global Trade Item Number data from bar code print data and allow you to simultaneously print and encode them into the RFID tag. The applications' flexibility allows you to select from many common shipping label templates such as ITF-14 and UCC/EAN-128.

Now that you can encode smart labels with or without EPC numbers, you have the means to test read ranges, read speeds and data capture. You can determine the distance from which the labels can be read, whether the products themselves affect RF signals, where you should locate the label on the carton, and variations to read angle and distance. As you become familiar with optimum read speeds and work out the intricacies of capturing and reading data, you will, most importantly, arrive at solutions to improving and maximizing system accuracy and efficiency.

Label Placement – Case analysis is the subject of next chapter. Package contents and label configuration, design, space and angle all can make a difference between a 100 percent read rate and a 0 percent read rate (Fig. 7.4). You will need to keep these factors in mind as you determine the placement of your smart label on the case or pallet.

 **SEE PAGE 207**

on case analysis.

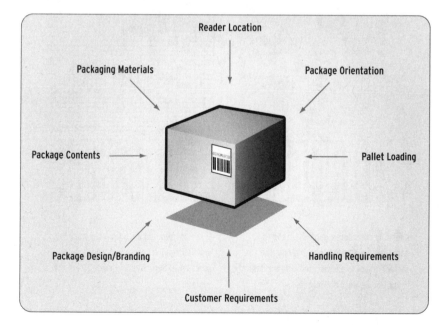

FIGURE 7.4

Label placement depends on a number of factors.

Detecting quiet labels – A label is considered good when the RFID data is written to the tag correctly, the correct image is printed, and content data is verified against the source. If the printed and encoded data can't be verified against the source, the label is considered defective and voided from the system (Fig. 7.5). To ensure that no EPC numbers are lost, the printer should be programmed to clearly overstrike and void the defective label and print another label using the same EPC data. When a verified tag can't be read from a normal

FIGURE 7.5

Examples of good and bad (quiet) labels.

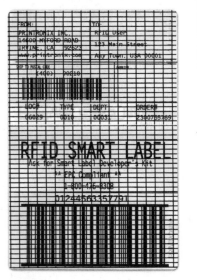

A label is considered good when the RFID data is written to the tag correctly, the correct image is printed, and content data is verified against the source.

If the printed and encoded data can't be verified against the source, the label is considered defective and voided from the system.

distance, it's called a quiet label. In some cases a quiet label may be the result of a defect in a specific label within a roll of good labels. Your print/encoding system should be designed to distinguish between quiet and non-quiet labels, removing yet another source of error. A quiet label needs to be eliminated from use if you want to achieve 100 percent read rates throughout your supply chain operations.

Simulate a dock door and a conveyor using fixed-mount readers. The Smart Label Developer's Kit creates a sampling of labels for typical cartons and pallets. (You might also choose to begin your testing in the established lab of an integrator partner for this initial phase.) Products such as Alien Technology's RFID Development Kit with reader, antenna and development system software will accelerate your progress through this phase.

Lessons coming out of numerous test lab experiences include:

LESSONS LEARNED

- One tag type does not fit all circumstances
- The least expensive tag may not be the best choice
- Media certified to interoperate with your printer/encoder usually provides the highest yields
- Standards and interoperability are important
- Evaluate what will be needed for rapid deployment
- Understand how to minimize deployment costs

PHASE 2: TEST AND VALIDATION

By now you should have a business case for one or more product lines and be ready to design and architect a solution. In this phase your team needs to engage knowledgeable partners to assist in testing and validating what will be the pilot implementation.

Here's what you should know as a result of phase 2 work:

- Label placement
- Sources for label stock
- Tag encoding and label printing strategy
- Strategies for meeting production volumes
- Physical layouts for the labeling operation
- Power and data network architecture for labeling stations and read portals
- Priority of opportunities for achieving internal business benefits
- System integration requirements to facilitate data flow

System Integration

→ SEE PAGE 249
on managing EPC data collection.

In this phase of test and validation, it is important to begin learning how RFID devices will integrate with your ERP (Enterprise Resource Planning) and WMS (Warehouse Management Systems). Testing and evaluation allow you to preview the extent of capabilities RFID brings to your enterprise and the supply chain. The volume of data coming from the reader network can be enormous. Evaluate the expected data storage requirements and network speed and bandwidth requirements. Get information on the scalability of middleware and WMS systems, so you are not caught short as you move forward.

RFID supports such areas of your business as resource planning, parts purchasing, order tracking, customer service, inventory management, transportation management and accounting. All of these areas will be affected by the integration effort.

Validate Vendor Choices

As you approach your pilot program implementation, evaluate how your equipment is working for you. No matter what manufacturer you team with, there are expectations you should maintain for RFID. For example, make sure your printer vendor offers:

- Complete encoding solutions (both manual and integrated)
- RFID extensions and drivers
- Support for EPC standards
- Ability to extend your development environment into production
- Certified smart labels in unlimited quantities
- Rapid development capability for your unique label design
- A modular and field upgradeable encoder platform

Printers and labels need to work together with the RFID equipment, communicating information back into the ERP or WMS system.

Reader vendors should be able to provide RFID solutions for various global frequency requirements as they evolve. During the implementation phase, readers can be positioned depending on your needs at various locations such as shipping and receiving dock doors, product routing conveyors, picking and sorting configurations, and forklifts. Handheld readers facilitate inventory counting, locating and reconciliation, and should have the ability to capture both bar code and RFID.

Point-to-point Testing

Conduct a deployment test from point to point in a controlled supply chain to form a deeper and more practical understanding of the capabilities and constraints of the technology, the impact on your organization, and the practical implementation issues to be addressed. You should now know the read ranges, read speeds and data capture capabilities of the basic system. You should verify your capability for generating an Advanced Shipping Notice and other shipping documents that combine case and pallet data. If possible, verify the demand triggering and event management capabilities that are tied to your customer's replenishment system.

PHASE 3: PILOT IMPLEMENTATION

The objective of the pilot program is to develop a predictable and scalable system. This requires you to achieve precision in placement, output and performance. Careful measurement and documentation throughout this phase will facilitate problem solving with your partners and selected customers, to ultimately eliminate errors and establish processes. You will want to mark critical milestones to chart the development of your system. Along the way, stop and assess your solutions – is smart label placement formulated and confirmed for different products? Should you run parallel pilots for different divisions of your business because of significant differences in processes? Is it time to incorporate additional readers and smart label printers to your system?

The pilot phase is the time to tool up for handling greater volumes with real-life criteria in actual working environments. You will build knowledge and confidence in the system as you work out the everyday demands faced by your business, even though you are applying the tests only to a limited volume. Figure 7.6 lists the steps for this and other phases. In order to achieve the pilot objectives, you will want to:

(→) SEE FIG. 7.6

- Set up equipment in other facilities/divisions to discover and solve any anomalies within each facility.

- Verify your ability to capture and transfer data and send it between locations.

- Capture data on each specific product out of a test run of assorted SKUs.

• Educate employees on the importance of the RFID system and how it will affect the way they work. If tags are being applied manually, this is a critical part of the learning curve.

• Partner with a retailer to send test shipments to verify system compatibility.

• Subject the system to the rigors of a typical production or shipping facility.

• Handle higher volumes (50,000 or more).

FIGURE 7.6

Steps for implementation.

RFID IMPLEMENTATION PROCESS

❶ GETTING STARTED	❷ TEST AND VALIDATE	❸ PILOT	❹ IMPLEMENTATION
Establish a team	Involve a knowledgeable systems integrator	Develop a predictable and scalable system	Explore opportunities for new efficiencies
Do a feasibility study			
	Evaluate/test various software applications	Set up equipment in other facilities divisions	
Assemble your lab			Capture and manage data
Set up a development environment	Evaluate/test with warehouse infrastructure	Verify your ability to capture and transfer data between locations	Implement RFID network and device management
Focus on technology and solution-based smart labels	Test read ranges, read speeds and data capture		Deploy smart media management
Select and prioritize target products	Conduct case analysis and select tag	Capture data by specific SKUs on a run of assorted SKUs	
Start making smart labels		Measure results	

- Measure results to test the viability of larger scale.

- Work with your partner team to eliminate errors.

- Consider expanding your pilots to additional products or geographies after successful completion of your first. You may find that different divisions or product lines require different pilots.

Solving RFID implementation issues, even if the requirement is for a small percentage of your shipments the first year, will provide a strong foundation for when 100 percent of shipments require smart labels. By the end of Phase 3, you will have locked down your business processes and procedures, tested software and hardware, and verified your system accuracy at higher volumes and speeds. You will have successfully accomplished your pilot implementation if you have:

- Measurement of results, including establishment of performance metrics.

- Integration of ERP/WMS to extract the data out of the label and pass the information back to the system for operations management.

- Defined different label and antenna requirements for different SKUs.

- Programmed your system to detect human errors in tasks such as label selection and placement as your equipment checks the data and media and alerts you to problems.

- Decided: manual or automated application of smart labels, in-process or post-process?

During 2005 and into 2006 only selected suppliers and only those products shipping to specified RFID-enabled Wal-Mart distribution centers are required to have full RFID compliance. It may be more cost effective to limit your RFID compliance to only those products shipping to those specific warehouses. If you are running out of time, consider the feasibility of making the application of smart labels a post-production step rather than an integral part of the manufacturing process. De-palletizing cases, adding smart labels and re-palletizing might be a viable short-term solution.

PHASE 4: IMPLEMENTATION

Whether you are just starting or already piloting, several issues considered in your early decision-making will facilitate a more efficient, successful implementation.

Ensure against obsolescence – Technology is evolving, standards are not resolved, and protocols will change. Smart labels are migrating to

UHF Gen 2. These changes in the industry will mean changes in your equipment. So choose a vendor offering asset protection to protect your investment, with free upgradeable firmware (for example, to support new standards or data formats) and scalable solutions so you won't have to start over. We suggest that you ask a set of pertinent questions to help you make informed decisions about product and vendor selection:

- How many pilots did they help deploy by the January 2005 deadline?

- Can they articulate examples and case histories from their experiences related to the Wal-Mart and DoD requirement?

- Do they have a strong list of RFID partners?

- Are they a global company to manage your international locations?

- Do the offer commercial EPC products in North America, Europe and Asia Pacific?

- Do they offer formally organized professional services such as label design and verification, on-site assessments, training, and integration and migration consulting?

- Have they already delivered additional protocol upgrades with free firmware upgrades?

During the implementation phase, you will explore opportunities for new efficiencies and build metrics into your processes to quantify improvements, forming a foundation for ROI. This reinforces that the solutions you pick for pilot runs need to be scalable, robust and industrial strength for cost-effective deployment. And even if your processes include manual application of labels for shipping at this point, it's important to keep your future automation capabilities in mind as your system expands. This is a critical factor when choosing your technology providers.

The Smart Label printing/encoding process is the first point at which an EPC number is entered into the supply chain. Any errors in this process will create significant problems down stream. Ensure that your printer/encoder has these characteristics.

Validation and verification – With validation built into printing equipment, you can correlate 100 percent bar code reads back to 100 percent RFID reads and be able to cross reference. Without manual intervention, your system will be able to check every label against your database to verify that what you read on the label is actually what it should read. If there is a discrepancy, it will immediately back up, cancel and overstrike the label, and print a correct replacement. Read-after-print quality control is designed to eliminate defective labels from entering the supply chain. It will also prevent print production slow-down, minimize the cost of labor, and avert product returns and fines for non-scannable labels.

Data capture – The ability to archive information for enterprise management brings the highest level of visibility to your operations. Through data capture, nearly instantaneous visibility of supply chain activity allows you to make more accurate sales projections and purchasing decisions. EPC data, once integrated into your database, can provide time, location, and batch information that when passed back into the system can identify and locate specific products at any point in the supply chain.

→ SEE PAGE 265
on reader administration.

Network and device management – Access to real-time information and control of your devices improve efficiency and productivity and help you make informed decisions in managing them. Network print management systems will provide instantaneous visibility to every discovered device and allow users to simultaneously configure an unlimited number of printers. These print management solutions will also support management of the additional RFID encoder capabilities. Providing instant visibility (enterprise view), instant notification through e-mail alerts and pages, and remote diagnosis, these tools will enable you to send test results over the printer network for viewing and storing in an XML file (or other formats) for later comparison with the data stream sent to the printer.

← SEE PAGE 134
on enterprise-wide print
management.

Smart media management – Your application will police itself and alert you if something is wrong. If the EPC doesn't correlate product to label, you will be notified immediately. Proactive detection will ensure the label placement is right, the class of label is right, and the antenna design is right for the label.

Industrial design – Your equipment is an investment in the future. To support you from test to pilot to deployment, your printer will need to be rugged and dependable, able to handle growing volume. Remember that the cost of a printer itself in the overall scheme of things is insignificant relative to the investment in infrastructure, tags and costs of non-conformance and non-compliance as you consider the downtime, lost productivity, fines and product returns.

VENDOR CHECKLIST

Consider the following requirements for your integration and vendor partners:

___ Participating member of EPCglobal

___ Partnered with other experienced RFID partners

___ Can provide field upgrade paths

___ Offers asset protection program

___ Upgradeable firmware

___ Solutions scalable to automation

___ Robust industrial equipment

___ Experienced in RFID specifically related to the Wal-Mart mandate

___ Deployed other pilots, can clearly articulate case study-type examples

___ Able to validate and verify labels without user intervention

___ Offer an enterprise network management solution

___ Formal professional services organization

___ Label verification testing, organize data and files for clients

___ Training, integration and implementation consultants

___ Global organizations to support clients' international locations and expansions

___ Able to supply volume quantities when needed

___ Equipment fully integrates with major supply chain software and other enterprise programs

___ Provides intelligent media detection avoiding the possibility of wasting expensive RFID labels on standard bar code print jobs

___ Offers RFID media certified to work with their equipment

Certification Programs

Radio Frequency Identification (RFID) technology has been in existence for more than a decade. Though its application has been limited for most of these years, it is currently one of the most promising and exciting technologies, thanks to its newfound and revolutionary applications such as supply chain management and asset tracking. Immense interest has been generated around RFID in the last three years.

The market is gradually waking to the fact that RFID technology is more than just the next step from bar codes or another data generating sensor-based technology. The onus is on industry participants to educate end-users on the basics, the challenges, the process requirements, standards and the returns from RFID.

Industry participants expect this market to grow explosively in the next 18-24 months. The growth of RFID technology market is expected to be driven by different factors such as mandates from retail houses, possible cost savings due to better visibility of inventory and governmental and/or regulatory body pressures.

Given such demand, according to CompTIA the industry is clamoring for RFID talent. A recent survey of CompTIA members found that 80 percent of companies do not believe there are sufficient numbers of RFID professionals in the work force today. In addition, 53 percent of companies said this lack of talent will have a negative impact on the adoption of RFID technology over the next two to three years. About 60,000 business will come under RFID usage mandates from trading partners over the next three to five years, according to CompTIA.

In response, CompTIA and UC Irvine have developed vendor neutral, foundation-level certification programs that qualifies and develops the RFID work force. CompTIA RFID certification is expected to be available in late 2005, early 2006.

In response to the rapid growth of RFID technology in such areas as supply chain management and product tracking, UC Irvine Extension is developing a new certificate program focusing on the implementation and use of RFID systems. The program is being designed to help business professionals understand the capabilities of RFID from a technical standpoint and from the perspective of improving a company's profitability by managing the movement of goods and products more efficiently. Having effective item tracking capabilities reduces the need for in-company storage and warehousing space, and ensures smoother transfers of goods and products from suppliers to customers. This certificate is designed for managers and analysts who wish to evaluate the ROI of RFID implementation and begin to devise an implementation strategy.

Contact Information:

Please visit the CompTIA website at www.comptia.org.

Contact Information:

UC Irvine Extension

P.O. Box 6050

Irvine, CA 92616-6050

http://unex.uci.edu

(949) 824-5414 Program Office

Stefano Stefan, RFID Certificate Program Manager

(949) 824-1367

Sources and Further Reference

Tag, Trace, and Transform: Launching Your RFID Program, white paper, Deloitte Consulting LLP, 2005.

"More on RFID from K-C and Wal-Mart," by Rick Lingle, *RFID Journal,* available: www.rfidjournal.com.

Allen-Bradley Distribution Center Projects Thousands in Labor Savings with RFID Implementation, Project Success Story, available: www.rockwell.com.

List of Printronix RFID Partners, available: www.printronix.com.

"Analysis of RFID Adoption and Workforce Issues in North America", Joint study completed by Front & Sullivan property of Front & Sullivan and CompTIA, 2005.

CHAPTER 8

Case Analysis

LAB TESTING 209

LOCATION TESTING 210

LABEL PLACEMENT 211

READER ANTENNA PLACEMENT 214

Tools for Case Analysis 216

CHARACTERISTICS OF RF AFFECTING READ RATE 219

IN THIS CHAPTER:

What to evaluate to determine smart label placement on packaging.

Case analysis is the process of evaluating packaging in order to determine where to place an RFID tag or smart label. The objective of case analysis is to achieve optimum tag readability in all the circumstances where the tag will be read. Case analysis should occur before you decide how the tag will be applied. Since you may be applying smart labels to cases by a variety of methods, it's best to determine what and where first.

A case analysis should answer these questions for you:

1. What needs to be done to the package to accept an RFID label?
2. Does the existing bar code label design and location need to be changed?
3. How will product composition affect RF signals?
4. What types of smart labels are needed?
5. What are the package label placement locations and tolerances for consistent RFID read rates?
6. What package material changes are needed to optimize RFID?
7. What is the palletizing strategy?
8. Will RFID work within my existing product packaging, palletizing and shipping process?
9. What training will be required for packaging, warehousing and distribution center workers?
10. How can I detect and minimize causes of RF interference?

11. Are there any asthetic or marketing requirements on the case that restricts label size, placement or appearance?

A survey done of tier one participants in the Wal-Mart compliance effort found that a good share of the problems experienced had to do with incorrect case analysis. In particular:

LESSONS LEARNED

- Incorrect selection of tag/label types for the products being shipped
- Incorrect positioning of labels, in some cases covering over existing important information
- Inconsistent positioning of labels, causing read variability
- Improper tag reading caused by encoding and read setup errors

LAB TESTING

The best place to start case analysis is off line, in a non-production environment such as a lab. In your lab you can start with a known good tag and reader, and determine how packages and package contents will affect RFID tag encoding and reading. You might want to consider the services of a systems integration company that offers RFID case analysis. They may have a facility already set up to conduct the lab testing phase.

Lab testing should be used to evaluate and test the following:

- Product chemistry, packaging types and composition, and how they affect RFID

- Tag and label choices, costs and availability
- Label placement options and tolerances
- Tag encoding and application methods
- Reader and antenna combinations
- Read rates at various line speeds and distances
- Pallet loading

LOCATION TESTING

Evaluation and testing continues as you move into the pilot and production phases. Questions that are answered during these phases include:

Divert or integrate? – Whether you create a separate compliance process and divert production to it, or integrate RFID into your standard packaging line, involves a number of factors. They include existing packaging and labeling methods, throughput rates, types and volumes of product that falls under a compliance mandate, read rates, internal returns gained by tagging upstream in the process, and investment timing.

RFID portals – The selection and setup of packaging line and dock portals involves an analysis of the physical environment within the reading area, the evaluation of various reader and antenna configurations, and performance testing. Antennas need to be tuned to operate within a certain range, and not overlap with the ranges of other antenna arrays which may cause interference. This is especially important in cross-docking applications. Some of these issues are being addressed by Gen 2 tags and readers, which can operate in dense reader mode and adjust themselves to noisy environments.

Short, medium and long term approaches – You should be able to determine the capability and capacity constraints of your initial approach, and begin planning for the medium and long term.

LABEL PLACEMENT

The location and orientation of a smart label on a case or pallet can be critical. Product composition, package geometry, packaging materials, pallet loading, proximity and orientation with respect to the reader antenna are all variables that have to be considered. In pilot applications for smart labels on packages containing liquids, label placement variation of as little as 0.25" (6 mm) from optimum has been found to affect read rates. As an example, some manufacturers of bottled drinks have found success by placing the tag at the bottle neck where there is an air gap. In the pharmaceutical industry, certain biological compounds used in medicine can change their molecular structure in response to RF. It is important to proceed within FDA guidelines if they apply.

LESSONS LEARNED

Tag presentation – As a tag passes through the read window, ideally it should be on the same plane as the antenna. The flat face of the tag should be parallel to the flat face of the antenna (Fig. 8.1). However, this has proven to be impractical and inconsistent with the mandates which require the case to be read from all six sides. Case analysis and placements need to take into account the ability to read in all orientations. See Figure 8.2. In cases where orientation is unpredictable, a tag with a dual dipole antenna may provide better read rates.

FIGURE 8.1

Ideal tag presentation is where the flat face of the tag is parallel and on the same plane as the flat face of the antenna.

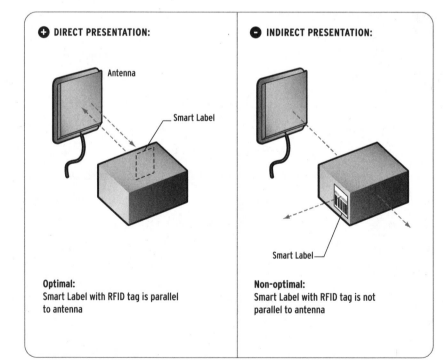

⊕ DIRECT PRESENTATION:

Antenna

Smart Label

Optimal:
Smart Label with RFID tag is parallel to antenna

⊖ INDIRECT PRESENTATION:

Smart Label

Non-optimal:
Smart Label with RFID tag is not parallel to antenna

Tag coupling – Tags placed on or very near metal objects, such as aluminum cans or foil, may electrically couple with them. This may short out the antenna. Proper tag selection and placement is very important with packages containing metals. Some smart labels are designed with a material layer that acts as a standoff to minimize the affects of metal or foil packaging. Air gaps designed into the case packaging can help minimize the interference caused by foil wrapped items inside.

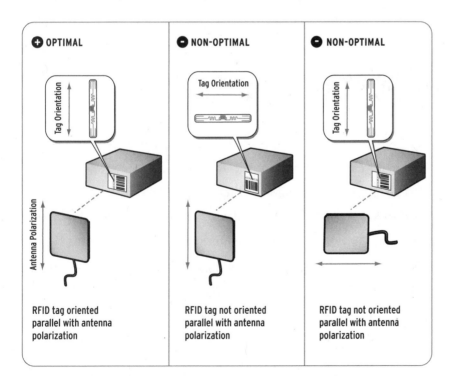

FIGURE 8.2

Label placement should align with antenna polarization.

Pallet testing – You may find that tags on interior or bottom cases on a pallet could become damaged and a new pallet loading strategy is necessary. An obvious aid to troubleshooting and recovery is having the EPC tag code printed on the smart label, in bar code and human readable form. By reading the printed EPC code, the individual cases with damaged tags can be identified. Current mandates do not require that case tags be read when the cases are palletized, only that the pallet tag be read, but mandates may change. You may also want to consider a software tool to automatically identify pallet loading issues and help find the optimum loading conditions.

⊖ **SEE PAGE 141**

on Wal-Mart system requirements and Page 171 on DoD requirements.

READER ANTENNA PLACEMENT

Since the power output of a reader is regulated and fixed, antenna design and placement is perhaps the most important way to tune the RF signal to an environment. Varying the reader antenna placement is usually the easiest troubleshooting step, but may be the trickiest one to do well. Figure 8.3 shows a typical dock door layout. Dock door portals can be coupled with a motion detector to turn them on. In cross-docking and replenishment applications it is important to establish the direction a pallet is going through a dock portal. Such a system will have two sets of antenna arrays aimed at opposite sides of the door, and signal logic to determine the direction.

⊖ **SEE FIG. 2.16**

on Page 65.

There are hundreds of different antenna designs for HF and UHF systems. Gain, efficiency and radiation pattern are variables that can be addressed by the design of a reading zone, and the selection of an antenna or antenna array. In general, antennas emit two types of energy patterns, linear polarization (Fig. 8.4) and circular polarization. A linearly polarized antenna may work best for a stacked pallet, but a circularly polarized antenna may be more appropriate for a conveyor application where the orientation of cases cannot be determined.

→ SEE FIG. 8.4

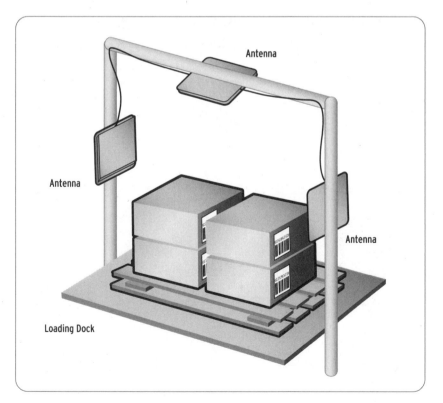

FIGURE 8.3
Dock door antenna placement.

Tools for Case Analysis

A useful tool for antenna selection and placement is a mapping of its radiation pattern within the operating environment. See Figure 8.5. This can be done by placing a known good tag or tag emulator at various points in the read area, and attempting to read the tag. By marking out a grid pattern on a floor diagram, indicating where the tag does and does not read, you can determine the practical range of the antenna. Since read redundancy (3 or more good reads within a short timeframe) is an important performance characteristic, especially if the reader or tag will be moving relative to each other, you should determine the limits of the read area.

FIGURE 8.4

Linear and circular polarization.

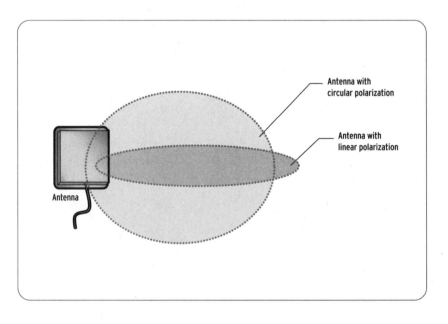

Antenna with circular polarization

Antenna with linear polarization

Antenna

Testing software and tools are available from tag manufacturers and RFID specialists to help map read areas and determine read redundancy.

Tag emulator or field probe – One such tool can be placed at various points in a reading area to verify reader range and operation. The tag emulator can also help determine the best tag location on a case. It not only emulates a tag, but also senses the interference and absorption characteristics of nearby materials. Several such probes placed on various parts of a pallet allows you to develop a

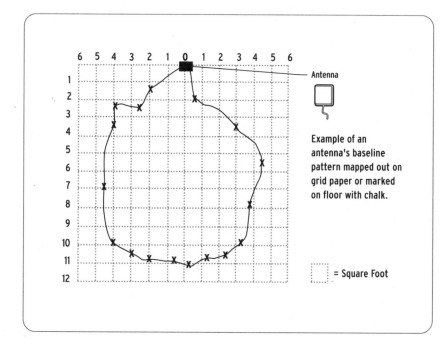

FIGURE 8.5

A floor diagram with tag reading areas boundaries marked.

3-D map of field strength and predict tag readability. The field probe can also assist in determining differences in the tag air interface implementations of various makes of tags and readers. Gen 2 readers, for example, have settings for the number of transmissions received from a tag to trigger a valid tag read. When combined with mapping software, a tag emulator will help in determining the read triggering threshold and the dwell time as a case moves through the read field.

Case and pallet analysis software – 3D imaging software can help a company design palletizing strategies that optimize RFID read capability. Integral RFID, a company based in Richland, Washington, provides a product called EPC Hot Spot, to help map RFID performance around a case or pallet. See Figure 8.6. The company also provides case analysis consulting expertise.

Other software in use by RFID integrators provides 3D visual representation of an entire pallet load, identifying all cases that read properly and those that don't. You can modify tag placement, case orientation and pallet geometry to test for and optimize reading of all internal and external tags on the pallet.

Auto-ID Labs Packaging Special Interest Group – This group is a source of extensive knowledge on materials analysis for RF tagging.

CHARACTERISTICS OF RF AFFECTING READ RATE

Passive-tag RFID is a new technology compared to bar codes. There are relatively few retail supply chain implementations, and few packaging engineers and system integrators with actual application experience. In contrast, many years have gone into the development of bar code printing techniques and the engineering of high speed packaging lines with bar code verification.

The hard science of RFID is just beginning to come to terms with the environment of a retail distribution center or packaging line. It

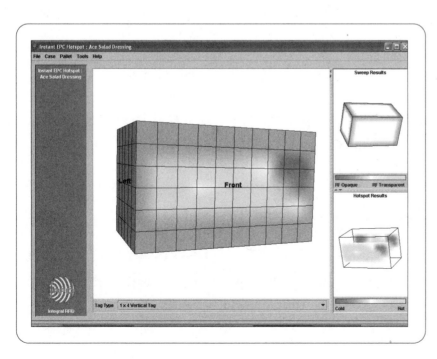

FIGURE 8.6

A 3D view of a case using the EPC Hot Spot software program.

is likely that every facility will be different enough to make the troubleshooting of read rate issues feel more like an art than a science, at least at first. Note that RF regulations in Europe present more of a challenge to read rates, because of limitations to frequency hopping and duty cycle. Similarly, first generation tag protocols that require a great deal of back and forth communication between the reader and the tag to establish a "handshake" will increase the chance of failure in a noisy environment. The Gen 2 protocol has refinements in signaling quality and speed that will improve read rates in most circumstances.

The following are characteristics unique to RF that affect read rates:

Translucence – Some materials offer little or no barrier to RF energy passing through them. Clothing made of organic and synthetic fibers, paper products, wood, non-conductive plastic and cardboard are translucent to RF. Paper packaging with foil lining, however, may block RF.

Absorption – Liquids, materials containing liquids such as foods, and liquids and foods containing salts in particular, will absorb UHF. Carbon containing compounds, such as graphite in solid or powder form, will also absorb UHF. What absorption does is attenuate, or weaken, the electromagnetic field propagating from a reader antenna or back from a tag antenna. Figure 8.7 is an example. Absorption varies by substance and by the frequency of a signal. It is possible to calculate the absorption rate of various substances to a certain frequency, and the resulting dielectric loss. Positioning tags in the air gap just below the bottle cap may reduce absoption.

LESSONS LEARNED

Shielding – Metals and very thin metal foils particularly can conduct a radio wave away from a target, not allowing it to pass through. Shielding material can behave like an induction coil, moving electrons in parallel with the induced current in a tag antenna, creating an opposing field that weakens the signal. See Figure 8.8. In general, higher radio frequencies are more easily shielded than lower frequencies.

⊙ SEE FIG. 8.8

Detuning – Tag antennas, are greatly affected by their immediate surroundings. A tag attached to a case of soda, for instance, is going to be more affected by its location (top of case, bottom of case, etc.)

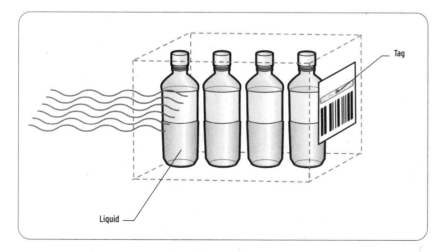

FIGURE 8.7

Liquids tend to absorb and weaken radio waves.

than anything else. Absorption and shielding from the cans will reduce the amount of energy that reaches the tag, and reduce the backscatter signal going back to the reader. Tags that are placed too close together can capacitively couple to one another, detuning their antennas. The metal on conveyors, forklifts and other handling equipment can also detune signals, by blocking and reflecting them. Tags with suitable antenna geometries, proper placement on individual cases, and proper case orientation on a pallet can improve read rates. Package re-engineering may also be required.

Reflection – At UHF frequencies, signal reflection is possibly the most important fundamental problem for RFID. Because of reflections, a reader signal may not penetrate a stretch-wrapped pallet,

FIGURE 8.8

Coiled wire could shield an RF tag by conducting energy away from it.

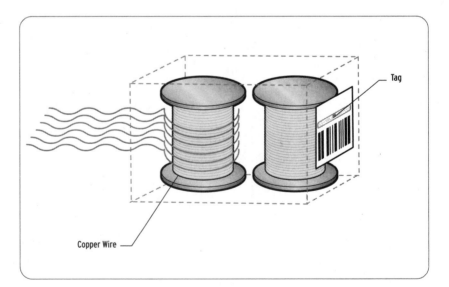

Tag

Copper Wire

for example, and the tags never receive enough energy to turn on. Metals reflect almost all of a radio signal. Some types of plastic films, coated glass and construction materials also bounce waves rather than let them pass. Reflections are due to the surface of a material having a different dielectric constant from that of the surrounding air (Fig. 8.9).

Interference – Interference creates so-called "dead zones" due to the geometry of the environment. Conveyor apparatus can induce dead zones through vibration or EM discharge from motors or controllers. Other RFID systems, wireless computers, radios, and phones all can create interference, but it is usually filtered out through the reader/tag air protocol. Electrostatic discharge, from

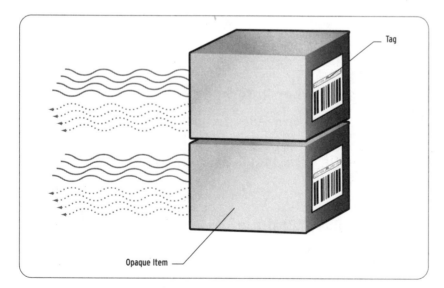

Tag

Opaque Item

FIGURE 8.9

Reflection of radio waves.

materials that accumulate static electricity and are not properly grounded, can also create interference. A reader signal can interfere with itself because of multiple reflections from other material surfaces. Examples include diffraction from surfaces as a signal goes through a narrow opening to reach a tag (Figure 8.10), or a signal that bounces off of a metal object and reaches a tag nearly simultaneously (Figure 8.11).

→ SEE FIG. 8.12

Achieving acceptable performance for a retail supply chain RFID application may require a multi-disciplined approach. Process performance analysis tools such as affinity diagrams (See Figure 8.12 for an example), Pareto analysis and design of experiments, may help identify the sources of problems. It may be necessary to repeat the case analysis activities with the help of a system integration partner.

Sources and Further Reference

RFID Primer, white paper, Alien Technology Corporation, 2002.

Wal-Mart's RFID Deployment – How is it Going?, white paper by Incucomm, 2005.

"Thinking Around all Sides of the Box," by Rich Fletcher, *RFID Product News,* Nov/Dec 2004.

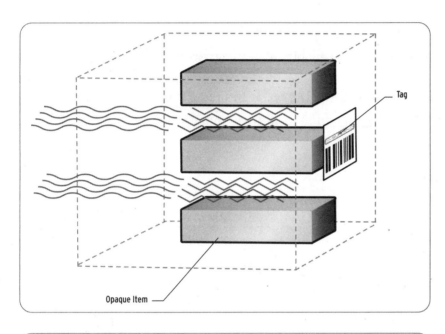

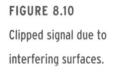

FIGURE 8.10

Clipped signal due to interfering surfaces.

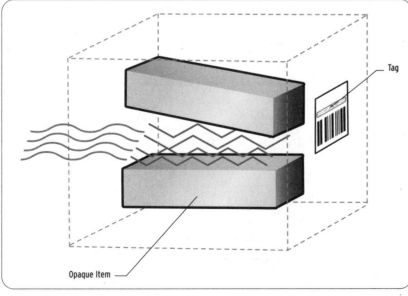

FIGURE 8.11

Reader signal interfering with itself.

Instant EPC Hot Spot case & pallet software, Integral RFID Company, Richland, WA, available: www.integralrfid.com.

Packaging and RFID Special Interest Group, available: http://web.mit.edu/auto-id/www/index.htm.

FIGURE 8.12

Example affinity diagram for troubleshooting read rate.

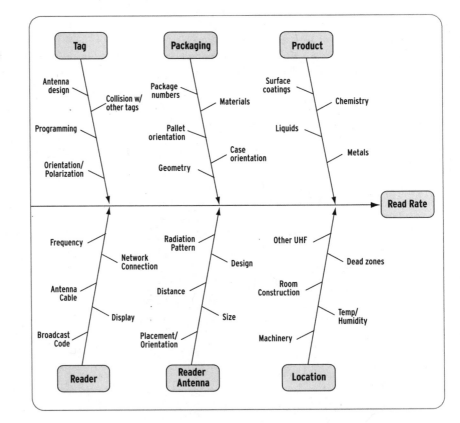

CHAPTER 9

Smart Labeling Approaches

MANUAL "SLAP & SHIP" 231

SLAP & SHIP WITH EPC MANAGEMENT 234

AUTOMATED "SLAP & SHIP" 237

Meeting Compliance Mandates With Slap & Ship 239

Smart Label Print & Apply Approaches 241

AUTOMATING MANUFACTURING PRODUCTION 244

SMART LABEL VALIDATION 245

A Pragmatic Path to Automation 246

IN THIS CHAPTER:

How to apply smart labels to meet compliance and production requirements.

Approaches to smart labeling are based on when a label needs to be applied, where on the case, and at what rate. Before any labeling approaches are to be considered, a case analysis should be conducted to determine the best tag placement to ensure successful reads. If your case analysis activity is successful, you have identified where to apply labels, and the best tag antenna and reader antenna configurations to use. Now you can look at employing one or more application approaches to meet either production or compliance requirements.

A fundamental principle of package engineering is to conform product to supply chain capabilities, not the other way around. Products come in all weights, shapes, sizes, composition and condition; packaging engineers put them in conforming shapes so that they can be handled with little exception by standard material handling machinery and transport methods. This streamlines the movement of product through the supply chain and guarantees that it reaches the customer in ideal condition. Smart labeling follows this same principle by conforming RFID to supply chain requirements.

The right approach depends on many factors. What may seem to be the best approach initially might upon further investigation turn out to be a wrong turn and a costly detour. Let's categorize and briefly review the approaches:

Tags integrated with packaging – Disposable corrugated packaging with built-in RFID has several attractive characteristics, most notably that it eliminates tag handling during packing and sealing. Encoding can occur before or after packing. Tag acquisition and integration costs can be pushed down to your suppliers. Disadvantages are many, however. Industry analysts predict that it will take at least 3 or 4 years for packaging companies to overcome the physical hurdles and make available a comprehensive offering. Error recovery, rework, and charge-backs may actually drive up costs. If you have multiple product lines that require different tag types and placement, you will have to purchase, inventory and manage a greater selection of package types, and match them with the product content. Not only will corrugated manufacturers have to invest heavily in dramatic manufacturing changes, but the delivery and supply chain for corrugated product will need to be redesigned. Today stacked corrugated is often left outside or in uncontrolled environments. The combination of heavy stacks, forklift abuse and hostile environments results in a waste percent unacceptable for product with embedded electronics. Lastly, paperboard packaging with embedded tags may have environmental impact and recycling issues that may be subject to green laws that can vary state to state and country to country.

Apply, then encode tags – Applying on-pitch adhesive-backed RFID tags before encoding is an alternative in instances where cases are pre-printed with a bar code. The apply, then encode process involves using an applicator upstream to affix the tag to a case, and an overhead reader to encode the tag as it moves through the packaging line. This approach could streamline the process and drive down costs, especially if existing applicator equipment can handle the tagging task. One drawback is that there is no way to validate or detect a quiet or bad tag prior to applying it to a case. This could cause a significant build-up of cases that require rework. Another disadvantage is inadvertent encoding of an adjacent tag that happens to be in the near field. Most label applicators in use today have not been designed to accept rolls of on-pitch tags. Tight turns and pinch points in some tag carrier paths could cause tags to lift and jam the applicator, causing severe damage to the tags. In addition, poor adhesion could cause tags to fall off with no clear visible indication of the tag being missing. The labeling industry has spent years perfecting adhesives for various applications; knowledge that many tag-only suppliers are just starting to acquire. As a result, this approach may add variability to the process, by increasing the potential for quiet or bad tags, unprotected tags getting damaged or lost from cases, and by potential encoding problems.

Encode, then apply tags – A way to address some of the shortcomings of apply then encode is to encode the on-pitch tags first, using an encoder built into the applicator. Unfortunately, approaches to this process are relatively unproven as of the date of this publication. On-pitch tag stock offers less flexibility and performance than smart label stock. In addition the tags by themselves are exposed and at risk of not adhering to the case because of limited choices for adhesives. Lastly, this approach does not always include the clear marking of a case with an EPCglobal logo to meet guidelines on RFID identification for consumers. Consumer notices pre-printed on containers could be subject to change as marking standards continue to evolve.

← SEE PAGE 106 on addressing consumer concerns.

Encode, print, then apply – Smart labels follow this approach, where an applicator is directed to encode and print a label on demand, just before applying it to a case. Validation, error-detection and recovery capabilities ensure that the tag is good, and that the application can occur at the appropriate point in the packaging process. The same assurances apply whether printing is done or not.

MANUAL "SLAP & SHIP"

Slap & ship is a term used to define a post-process tagging procedure. It can be divided into 5 key stages: 1) identify and sort product to be tagged, 2) de-palletizing the cases, 3) encode and apply smart label, 4) verify and transmit EPC, 5) remark the pallet. The encoding and applying of smart labels can be either a manual or automated procedure within slap & ship. Manual slap & ship is applying a label by hand to a case or pallet, usually after the production process. Manual slap & ship may work as a first step for some users to apply tags to low volume, RF friendly product, but will eventually reach a tipping point where costs and productivity require a shift to automating the process. Manual slap & ship can be labor intensive and slow. It runs counter to the long-term goals of the speed and labor savings vision of RFID technology. Slap & ship is where some RFID applications start, however, in the lab and in early pilots, to verify placement, read rates, read speeds and other issues. A compliance labeling application can be initially set up as manual slap & ship with EPC management, as described in the next section, then converted to an automated slap & ship with the addition of an integrated RFID printer/applicator.

Figure 9.1 shows all of the locations where manual on-demand labeling may have value. It is appropriate for post-processing operations like de-palletization of cases to be reworked for RFID compliance, for reject lines, returns, special handling, or where the RFID requirement does not justify a full production line implementation.

→ SEE TABLE 9.1

Table 9.1 compares some of the pros and cons of slap & ship. Most distribution centers set up for RFID compliance will need a slap & ship capability for on-demand encoding and printing of labels.

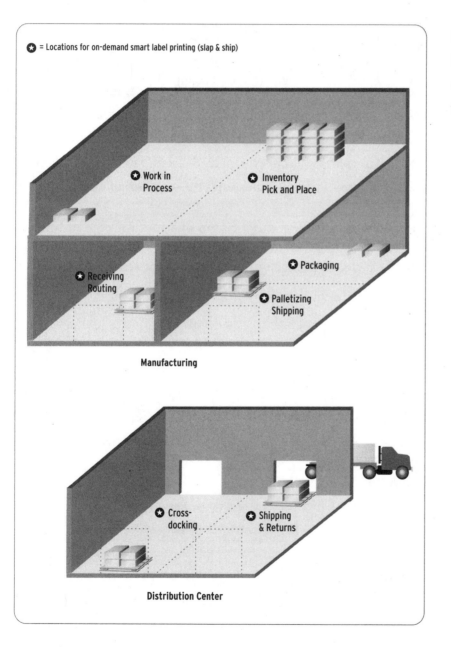

FIGURE 9.1

Locations for manual on-demand labeling for RFID.

SLAP & SHIP WITH EPC MANAGEMENT

A slap & ship approach can help a company get started with compliance labeling in a productive way. Products to be shipped with RFID labels would be separated and diverted to a staging area, particularly if production line speeds prohibit slap & ship. The diversion can be achieved by splitting a conveyor, or by delivering individual pallets to the staging area. Figure 9.2 illustrates the sequence of operation:

1. Cases are de-palletized.

2. The bar code for each unique case type is scanned. Product is identified with labeling and shipping requirements.

TABLE 9.1

Advantages and disadvantages of slap & ship.

Advantages	Disadvantages
Most flexible approach	Labor intensive
Least engineering content & cost	Limited integration with process
Allows tagging at distribution center just before shipment	Does not provide upstream benefits of RFID
Fast start-up	Low volume
Back-up for products tagged on manufacturing line	Does not scale
Achieves initial compliance labeling requirements	Lengthens investment period
Potentially least disruptive approach	May require separate staging area
On-demand	Potential for operator error
Allows company to learn about RFID	Long-term competitive disadvantage

FIGURE 9.2 Sequence of operation.

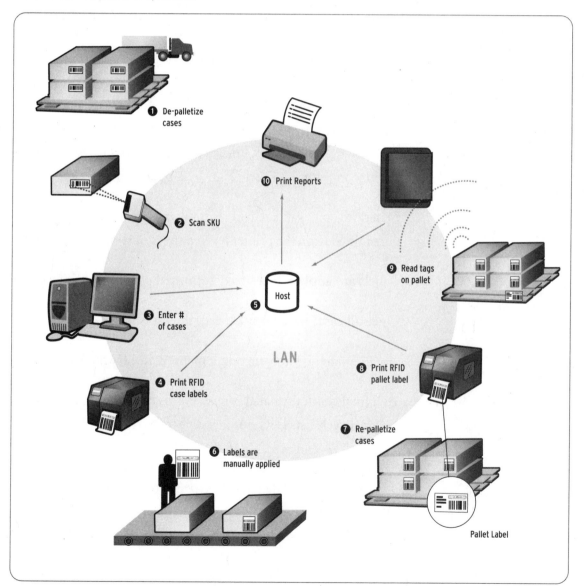

1 De-palletize cases

2 Scan SKU

3 Enter # of cases

4 Print RFID case labels

5 Host

6 Labels are manually applied

7 Re-palletize cases

8 Print RFID pallet label

9 Read tags on pallet

10 Print Reports

LAN

Pallet Label

3. The number of cases is entered into a computer. A series of unique EPC codes are generated and their usage is recorded in a local database.

4. Smart labels for each case are encoded and printed with unique EPC codes. The smart label RFID printer can be positioned within easy reach of workers by putting it on a mobile platform with a wireless network link. It can also be connected directly to the computer by Ethernet or by a parallel interface.

5. Generated and validated EPC numbers can be confirmed and transmitted from the printer back to the host system for tracking.

6. Labels are applied to a pre-determined position on each case.

7. Once all cases have been labeled, the cases are re-palletized.

8. A pallet label is then encoded, printed and manually applied.

9. As a final check the pallet is read and compared with the pallet's original pick list. EPC numbers are confirmed with shipment and ASN is generated.

10. At the end of each day management reports can be printed.

AUTOMATED "SLAP & SHIP"

The application of a smart label is done at the point where the process knows that the pallet is destined for Wal-Mart (or another customer having an RFID initiative). Automated compliance labeling can be done at a DC or outsourced to a third party logistics provider. A rationale for this approach is that a company wants to minimize the impact to its total operation since only a fraction of product goes to that customer, but the volume of product or it's content makes manual slap & ship impractical.

Figure 9.3 illustrates an automated compliance labeling approach. The majority of semi-automated approaches would use a conveyor system in conjunction with a label applicator integrated with a smart label printer. The conveyor is fed manually with cases from a de-palletizing station. A separate smart label printer is stationed at the end of the line. Once all cases are tagged, they are manually or automatically re-palletized and stretch-wrapped. A pallet label is encoded and applied before the pallet exits through an RFID portal.

→ SEE FIG. 9.3

The benefits of this approach include:
- Achieve precision of label placement
- Achieve speed of automation for volume SKUs
- Reduced manual labor
- No interruption of shipments to other retail customers
- Controlled investments that support preparation for future needs

- Focus on identifying and solving RFID technical challenges
- Methods would be applicable to other compliance requirements
- Integration with existing WMS and ERP
- Supports mixed pallet shipments
- Fully EPC compliant

FIGURE 9.3 Automated compliance labeling

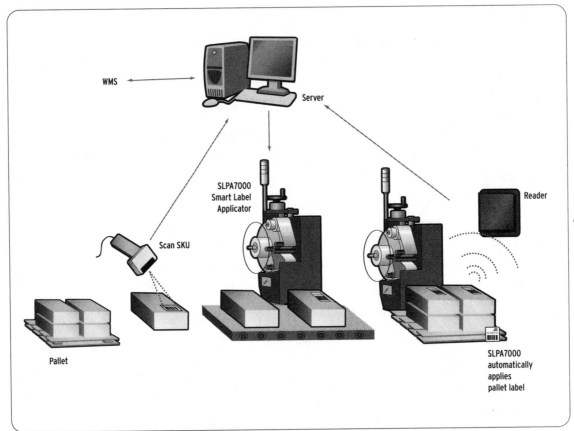

There are two types of print & apply designs: loose loop and next-out. The loose loop mounts the printer on a standard label applicator. The printing and applying occur at two separate locations. A queue of printed labels – normally 10 to 20 labels depending on label size – exists between the printer and applicator. A next-out label printer applicator is an integrated piece of equipment with no printed label queue. The printed label is immediately available to apply. The label printer applicator is preferred when the label data changes randomly or with each label.

Figure 9.4 shows an example of a label printer applicator. Error recovery is built into this approach. If a quiet tag happens to be on the smart label roll, it is detected automatically. The Printronix SLPA7000, for example, keeps the label with the quiet or defective tag on the label liner and automatically advances to the next tag. No reject label holding area required – which is an extra operation. The rejects are rewound on the liner thereby facilitating collection and removal. A record is kept of the number of quiet and no-read tags. The applicator can either apply the next good tag with duplicate information or sequence to the next EPC number.

→ SEE FIG. 9.4

Meeting Compliance Mandates with Slap & Ship

If you look at your RFID adoption curve as a conservative 7-10 year timeline, a slap & ship approach may be in use in various parts of your operation for a number of years. The time to transition from slap &

ship to a more automated approach depends on a number of factors:

Product portfolio – High value items subject to lots of shrinkage justify the investment in RFID, and are immediate candidates for an automated approach. Products and packaging that are not "RF friendly" and require package and process engineering to achieve compliance may take longer to transition away from slap & ship.

FIGURE 9.4

Label applicator with integrated RFID printer.

Technology maturity & reliability – Automation requires a heavy initial investment and a longer period to re-coup costs. Automation usually entails building more rigidity into a process in order to achieve reliability. Unless you have at least 100 boxes coming down a line that need tagging, manual tagging can suffice. Since RFID standards and technology are continuing to evolve and change, there is risk in committing to a more automated approach too soon in the RFID adoption curve.

Competitive landscape – What your competitors do to meet RFID mandates will affect your decision when to automate. It will be important to acquire internal as well as a broad industry understanding of best practices. You should be developing a business case for automation well in advance of having to react to a competitor.

The tipping point, when you automate the application of RFID tags, is different for every company and application. The point can only be determined when you've evaluated the merits of slap & ship and print & apply approaches at various case volumes and points on the adoption curve.

Smart Label Print & Apply Approaches

Smart label print & apply is the term used for an automated print, encode and apply labeling process. Smart label print & apply requires

← SEE FIG. 9.4

a label applicator integrated with an RFID printer which can be synchronized with cases moving on a conveyor (Fig. 9.4).

Considerations for choosing an automated approach include:

Case Volumes – The number of cases to label per week or per month should have a significant impact on the decision to automate.

Yield – An applicator is faster and more repeatable in its label placement accuracy.

Application – Application methods vary, and will change with production requirements. For example, label registration tolerances for RF hostile products may require a tamp-blow application to insure tight tag placement tolerances.

→ SEE PAGE 317
for case studies.

Labor cost savings – An applicator provides an upstream solution which translates to reduced labor costs, especially when read curtains, sensors or bar code scanners are used to trigger downstream events that build internal efficiencies.

Lead Times – Time sensitive products may require users to apply tags within a narrow window of time to meet customer deliveries. In such cases, users may require an automated solution to satisfy customer delivery requirements.

Tag Quality Management – Since converters can't always guarantee 100 percent good tags, an applicator has to have a way to handle rejects, dispensing with the bad tag and making a duplicate that can be applied. Reject handling slows down the operation and must be engineered into the system. Some applicators discard defective tags into a bin or onto a lever arm that fills up and stops the machine, requiring operator intervention. This approach prevents users from performing a post-mortem on defective tags (a process many companies currently use as a quality measure of their system). Since this process creates a "label-brick", users cannot analyze defective tags. A better approach is to detect defective tags before they are separated from the carrier liner, and leave them to accumulate on the take-up reel. This also makes it easier to account for defective labels and address yield issues with the converter.

Results from case analysis – Some products which are not RF friendly will require very precise label placement, not achievable with a manual apply process.

ESD – Smart labels are susceptible to damage from electrostatic discharge. Electronic component assembly techniques, including proper grounding of equipment, must be included. A typical warehouse operation can generate 8-15 kV of electrical discharge according to ISO estimates. It takes as little as 500 V to damage a chip.

AUTOMATING MANUFACTURING PRODUCTION

→ SEE PAGE 324
on Case 2.

Companies may consider splitting lines and running parallel operations to increase output. A tandem set of applicators, or several in a series, synchronized through a packaging line controller or other system, can double or triple the yield, and provide more flexibility on mixed-use packaging lines.

Smart label print & apply units can interface with practically any conveyor system via software, hardware programmable logic controllers (PLC) or in some cases, their own GPIO (General Purpose Input/Output) boards. These logic controllers, through an input/output module, sequence the smart labeling process with upstream readers, conveyors, light stacks, alarms, and other equipment. Light stacks alert operators of certain events such as: label rolls running low, no labels on tamp pad, no RFID media, etc.

At the secondary (case) packaging level, production line speeds usually vary from 20-100 cases per minute. Some smart label printer applicators can now operate between 20-50 cases per minute. To achieve higher speeds to address higher through-put requirements may require two or more encode, print, and apply devices.

SMART LABEL VALIDATION

The printing and encoding of smart labels needs to be a completely error-free operation. The smart label printer needs to recognize the occasional faulty RFID tag and prevent it from being used on a package. It also needs to provide validation data to a host computer for traceability.

A typical high-volume distribution center may print and scan 50,000 to 100,000 routing labels a day. An error rate of 3 percent can cost more than $1 million a year in rework labor expense. Since Wal-Mart is requiring 100 percent read rates for smart labels, and proof of validation, a smart label printer with EPC data validation plays an important part.

Validation Management – Integrated with a host computer, a smart label printer operates a closed loop system to validate its operation. The smart label is validated against test parameters. If the label meets specifications, it is ready for use. If the label does not meet specifications, the label is voided, marked "bad," and a replacement label is encoded and validated. The printer also analyzes the incoming data stream and for each label associates the tag content programming commands with the tag content read after encoding.

The validation activity is managed at the printer without host computer intervention. What is passed to the host is a validation log, indicating the number of labels printed, number canceled, and statistics by type of failure. By capturing this data and making it available to the host computer, the RFID printer closes the loop so that the host is aware of all successful tags encoded.

A Pragmatic Path to Automation

A pragmatic approach requires an assessment of your product line using these rough categories:

RF friendly – Begin applying automation to SKUs with the highest internal ROI and the least sensitive content. RF friendly products are those that do not have high absorption or reflective qualities (liquids and metals). If they are packaged in paperboard material, and are flexible to tag choice and location. A case analysis should indicate that you can get good reads regardless of orientation, even when loaded on pallets.

RF neutral – Cases that have dense product composition will exhibit moderate sensitivity to tag choice and location. They may be good picks because of their high value compared to labeling cost, and relative ease of labeling.

RF sensitive – Cases containing liquids and metals, or products in foil lined packages are RF sensitive. Air spaces between product can make a case somewhat RF lucent. Nonetheless, a case analysis will probably indicate labeling is consistent only within tight tolerances, with less-dense pallet loads. RF sensitivity may contribute to higher labeling costs as you move upstream to automate the process.

RF hostile – Cases that are opaque because they are densely packed, and contain products that attenuate the RF signal, will probably be the most expensive to label, and may be the biggest challenge to achieving ROI by applying upstream automation.

A business assessment should follow the product assessment prioritization. A decision to go forward should include an evaluation of available technology, system integration and process transformation requirements.

Sources and Further Reference

Section on "In-line RFID Print & Apply" contains numerous contributions by Rick Fox, President, Fox IV Technologies, Inc.

"RFID Technology Delivers Smart – but Sensitive – Packaging Labels," by Brian Albright, *Frontline Solutions,* November/December, 2004.

"RFID print and ship," by Bob Trebilcock, *Modern Materials Handling,* January, 2005.

CHAPTER 10

Managing EPC Data Collection

APPLICATION INFRASTRUCTURE 250

Components of the RFID Network 252

Regulatory and Operational Requirements 253

Network Traffic 256

HOST TO READER COMMUNICATIONS 257

Gen 2 Host to Reader Interface 259

Application Level Events (ALE) Specification 261

Event Management 264

Reader Administration 265

Security and Quality of Service (QoS) 267

GLOBAL EXCHANGE OF SUPPLY CHAIN DATA 268

GLOBAL DATA SYNCHRONIZATION 271

IN THIS CHAPTER:
How readers seek
data from tags and
communicate with
host computers.

The major benefits of RFID come from wireless computer-to-tag communication, but only if the data exchange is made useful. Through RFID, computers can identify, track, document and direct the movement of items through the supply chain with no human intervention. Data integration is a bottom up exercise. The groundwork involves the smart labeling of cases and pallets, and installing readers at key touch points in the delivery cycle. The next step is the integration of RFID with control and information systems.

APPLICATION INFRASTRUCTURE

A layer of applications support and collect information from every RFID implementation (Fig. 10.1):

Corporate planning – At the top of the information model are the human, financial, procurement, customer relations and product development resources of a company. The corporate planning system would be the keeper of the product catalog and ordering system, which includes product order numbers, SKUs, trade item identification, customer information and delivery requirements.

Manufacturing and distribution – A master schedule sequences product manufacturing, packaging and shipping. A forecasting system may drive case builds and inventory in anticipation of receiving the actual customer order. Product and order information converge at the point where the actual labeling and order fulfillment takes place. This may be at the end of a packaging line or

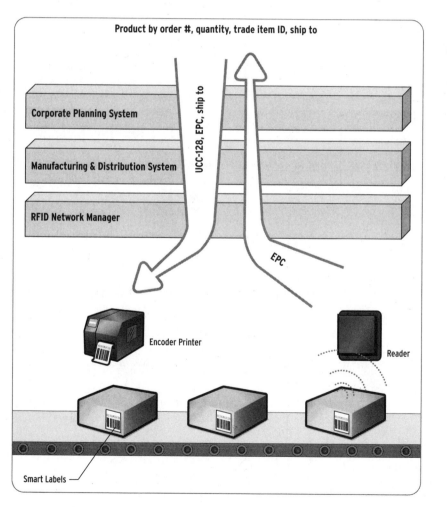

FIGURE 10.1

Operational data flow.

at a warehousing and distribution center.

RFID Network Manager – This layer shields the upper-level applications from the flood of data produced by a reader network, and provides a standard way to translate the business logic of an application into device commands and data collection routines. One side of the network deals with commands to program and apply tags. The other side manages the collection and dissemination of data. If smart labels are being used, tag programming becomes part of the printer/encoder job stream. An enterprise print manager translates orders to print and apply labels under the direction of the manufacturing and distribution system. EPC numbers may be created here, or passed down from another layer.

Components of the RFID Network

The basic components of the RFID network can be assembled in various configurations and topologies suitable for a distribution center, a product manufacturing and packaging line, retail outlet and other operations. These networks would link to each other through the upper layers of the information model. The components include:

Tags – The generic term for a radio frequency identification device.

Printer/encoder – Produces programmed tags as a stand alone station or integrated with an applicator.

Reader – Depending on the model and application requirements, a reader can be a relatively sophisticated and autonomous device, with multiple sensors, database memory and event triggering. Or, a reader can be a relatively simple listening device, integrated with and dependent on other network devices and host computer software for its intelligence.

Network – The network can be wireless or wired. It needs to be designed for transaction speed and volume, and integrate with other layers for communications and device management.

Data collection and device management software – Software is needed to manage the RFID reader arrays, filter redundant reads, compile lists of tags by various attribute, and pass EPC data to the manufacturing and distribution layer.

Regulatory and Operational Requirements

Most RFID reader networks can be put together with off-the-shelf LAN hardware and an Ethernet or wireless connection to a server. A reader network, however, is not like a typical office LAN that handles email, file sharing and print jobs. In some ways it is like an airline reservation system, where high speed and frequent transaction processing drives events. In other ways it is like an asset tracking system, where financial measures and regulatory rules are at stake.

An RFID network is tracking tangible assets – balance sheet items – and tracking them accurately has broad implications. Table 10.1 lists potential regulatory issues. Agricultural, food, drug and beverage companies that import into the USA, for example, fall under the Bioterrorism Act, which requires product tracing and tracking and regulatory data management. Approximately 200,000 facilities in over 100 nations have registered with the United States public health protection agency, the FDA, as required by the Bioterrorism Act. Regulators in the US and other countries are demanding that manufacturers have a way to trace the origin of each and every ingredient in a recipe, including details of who processed it, when, where and how.

TABLE 10.1

Regulations related to data collection and reporting.

Regulation	RFID Related Requirements	Organizations Affected	Governing Body	Reference Link
Bioterrorism Act of 2002	Dec. 2004 ruling on establishment and maintenance of records	Importers of food for human and animal consumption	U.S. Food and Drug Administration	www.fda.gov
Customs-Trade Partnership Against Terrorism (C-TPAT)	March 2005 incentives for "Green Lane" shipping clearance for smart tagging	Global supply chain logistics providers and importers	U.S. Customs and Border Protection	www.cbp.gov
Sarbanes-Oxley	Section 404 on integrity and visibility of financial records	Public Companies	U.S. Securities and Exchange Commission	www.sec.gov

The C-TPAT program affects all importers. Since the so-called Patriot Act of 2001 made it a federal crime to knowingly falsify a shipping manifest, the onus is increasingly on importers to meet regulations aimed at container cargo security. The C-TPAT guidelines provide ways to help importers achieve "green lane" status for shipments, using advanced ship notice and active RFID tags to electronically seal container loads. Eventually, a nested electronic manifest consisting of passive tag data from case/pallet contents will be tied to the active RFID seal. Using these measures, global supply chains can be more secure and transit clearance more predictable.

The Sarbanes-Oxley Act for financial reporting, record retention and disclosure, requires that networks be designed to pass an audit for data integrity. SOX policies link raw and finished goods inventory records, which are maintained by WMS and ERP systems, to financial statements to which the CEO and CFO are held accountable.

Beyond the regulatory issues are operational issues related to supplier mandates, labeling and ASN procedures, and the goal of gaining efficiencies. Dedicated networks for identification data may become the norm in a fast-moving retail goods environment, once a company goes beyond the pilot stage, and begins an ROI-focused implementation that links to a supply chain. Careful planning should go into designing the network, and selecting components for performance, compatibility, resiliency and ease of administration.

Network Traffic

Most pilot implementations have low volume. According to an IDC study, a full scale implementation will need to factor the following into its volume calculations:

The number of tags in circulation – Your master catalog of SKUs and procurement parts list are the starting point if you are going to size a network for item level tagging and for receiving raw materials from your own supply chain.

The number of times a tag is scanned – A computer manufacturer, using bar coding in its assembly operation, for instance, reads each product over 250 times as it goes from station to station. Each scan generates an automated event, such as a stage complete notification or alert. When an RFID reader senses a tag within its read area, it can repeatedly poll the tag as it moves along. Given that Gen 2 tags can transmit at up to 640 kbps, a reader can potentially collect a mass of data about just one tag.

The amount of data generated each time a tag is read – A typical RFID tag will have 128-256 bits of information, within which as a unique 64-96 bit EPC code. Readers can be configured to simply pass tag data to a management layer, expecting the host to filter and interpret the data. Or, a reader can heavily filter the data, passing a limited amount of data over a network.

Most systems will have a blend of device-level intelligence and management software intelligence which needs to be factored into the design because it impacts network data volume. Some readers can be configured to look like a SQL database and HTTP server on the network. They respond to a structured query command, but otherwise stay silent, accumulating tag data in their buffers.

The traffic volume generated by tag data continues if you are going to use EPC numbers as license plates for looking up data in company or remote databases. It is possible for tags to trigger calls to remote product information IP addresses listing a wide range of details, such as the product's manufacturer, customer, delivery, price, expiry date and more.

HOST TO READER COMMUNICATIONS

For most applications, readers will operate in one of two ways, either autonomously or as directed/interactive devices. Anti-collision algorithms, combined with a sequence of scroll, quiet and talk commands, are used to read and sort multiple simultaneous incoming tag signals.

Autonomous mode – A reader can be set to continuously operate, accumulating lists of tags in its memory. Tag lists represent a dynamic picture of the current tag population in its read window.

See Figure 10.2. As tags respond to reader broadcasts, they are put on the list. If they don't respond they are dropped from the most recent list stored in memory list. A persist time is set to determine the duration between the time a tag was last read and when it is removed from the list. A host system on the network can receive a list of tags from the reader whenever it chooses to listen. The information available to the host would include the reader location, time read, the size of the tag list, and the IDs of the tags on the list.

Directed/interactive mode – Readers in this mode will respond to commands from the network host. The host can instruct the reader to gather a list of tags within its read window, or look for a specific tag. In both cases the reader starts by gathering a list. Once it completes the host command, the reader waits until it receives another.

FIGURE 10.2

A reader set to read smart labels autonomously.

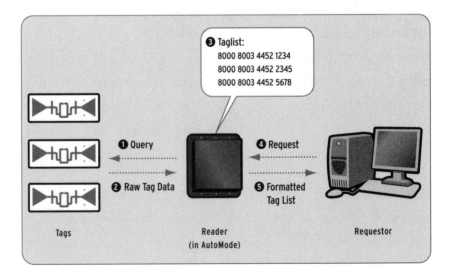

❸ Taglist:
8000 8003 4452 1234
8000 8003 4452 2345
8000 8003 4452 5678

❶ Query

❷ Raw Tag Data

❹ Request

❺ Formatted Tag List

Tags

Reader (in AutoMode)

Requestor

Gen 2 Host to Reader Interface

The host to reader interface specification for Gen 2 is a harmonized set of minimum standards. The standards were written to allow existing Class 0 and Class 1 compatible readers to become compatible with Gen 2 with a firmware upgrade. It is likely, however, that the more powerful features (dense reader mode, 640 kps data rate) will be available only in new reader designs.

A conceptual view of a Gen 2 reader is shown in Figure 10.3. The view is divided into read and event subsystems, where the read subsystem is configured to validate a good tag, and the event subsystem

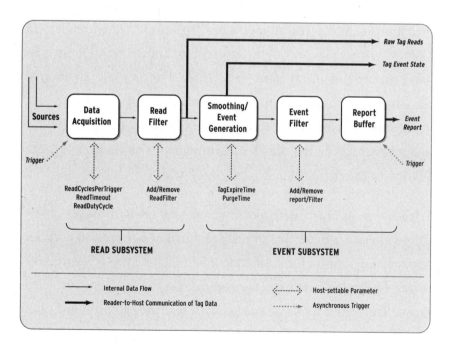

FIGURE 10.3

Conceptual view of a reader.

to manage the communications. Characteristics of the Gen 2 interface include:

Read source definition – A read source could be an individual antenna, antenna array, bar code reader or other sensor.

Virtual reader – A single physical reader can be divided into several "virtual readers" each assigned to a separate antenna. This allows one physical device to be installed, say at a dock door, and assigned one network address, but able to control several read zones (dock side and warehouse side of the door, for example).

Filters – Readers can be set with filters to ignore tags that do not match a preamble. This allows a host to set individual readers to look for certain tags, or only communicate known tags. Filters help make the volume of tag data manageable. For example, a personal computer system might contain a number of tagged components, the keyboard, hard drives, monitor, etc. The reader at a dock door might be set up to filter out all sub-components and collect only the STGIN on the case.

Trigger – Triggers are used to define a tag read event. A trigger can be a time interval, a pulse rate of reads, a signal or a transition event. A minimum definition for a valid tag might include two trigger events corresponding to when the tag entered the read area and when it exited. The reader might be sampling its read area continuously, but not place tag data into its read buffer until the tag exits the read area, or gets three valid reads. On a packaging line, for example, a trigger to read the package could come from a sensor on the conveyor.

Event filter – These define what tags are admitted to the read buffer. The "history" of the tag as it traveled through the read area (time read, how many reads, time left, associations with triggers, etc.) along with the filtered tag data can be used in the definition.

Application Level Events (ALE) Specification

Early adoptors of passive RFID for supply chain realized right away that tags and readers produce huge volumes of data that would overload business applications with a flood of data. Middleware, originally called "Savant" by the Auto-ID Center, is needed to collect, filter and add context to reader data, thereby providing business applications with useful information rather than raw data. More than sixty companies collaborated in an EPCglobal working group to define a specification to meet that need. The resulting version 1.0 ALE specification is currently pending EPCglobal ratification.

ALE provides a device independent application programming interface (API) for reader networks. Once adopted and made available in middleware components by software developers, it will provide a common language which business applications can use to talk to readers. ALE insulates the business application from the specifics of the RFID data sources, whether they are wireless handheld readers or dock door readers connected to antenna arrays. By providing this layer of abstraction, RFID networks can evolve and grow without causing ripples of reconfiguration work in the applications.

ALE middleware serves to collect, count and filter data for business applications. See Figure 10.4. Presummably, device vendors will embed ALE functionality into some reader models, and application vendors may provide an ALE plug-in, possibly eliminating the need for an actual third piece of software.

FIGURE 10.4

Logical view of an ALE compliant network.

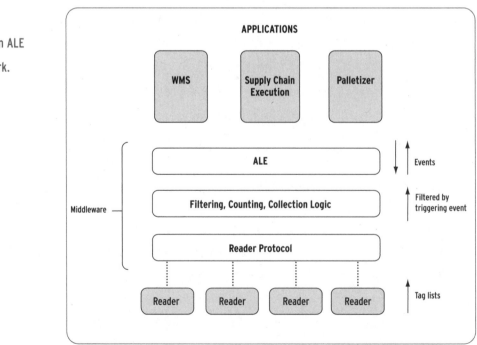

The ALE specification contains the concept of a physical and logical reader. A logical reader can consist of one or more physical readers. For instance a dock door portal, with both inside and outside read areas, can be classified as one logical reader. ALE is independent of the Gen 2 specification. ALE middleware will interpret ALE commands into the appropriate host-to-reader interface, shielding the business application.

For a given set of readers and associated devices, the ALE-compliant management system provides:

- Address mappings of read locations
- Rules as to how to accumulate data, by time, by triggering event or by command
- A library of filters to apply to the data
- Data aggregation rules so tag lists can be grouped by SKU, vendor, manufacturing step, etc.
- Tag counts as well as lists of EPC numbers

In effect, ALE turns RFID data into transaction events, which a business application can easily interpret and manage. ALE facilitates repurposing and sharing of data, allowing a set of applications to sense and repond to events.

The ALE specification also provides a way to configure a network to look for and record data from only certain types of tags. You can do global searches of tags for alerts and recalls. For example, a search for a bad batch of aspirin can be initiated through the ALE manager, to survey all read locations and locate the batch, without interrupting normal data collection activity.

Event Management

By turning tag reading activity into transactions, ALE software provides information which upstream applications can use to interpret and manage events. For example, a series of transactions go into building an ASN, including a case count, the EPCs of each case, the associated pallet EPC, and shipping information. Using the ALE command set, the WMS can run queries to gather this data, format the ASN and initiate the ASN transmission at the appropriate time. Similarly, a pallet unloaded from a truck and read at a dock door portal is interpreted as an event that triggers a cross-docking workflow.

Every tag read and acknowledged by a system implies an event. Typical responses to each event would include a database update, a trigger to perform the next sequence, or a trigger for an exception-handling routine. The WMS uses the last known position of the tag to infer the next step.

Reader Administration

A network consisting of more than a few readers will benefit from a software package that remotely administers the devices. Most readers can accept configuration commands via SNMP (simple network management protocol), providing a standard method for centralized control. Since a Gen 2 reader chip-set is in effect a software-defined radio, a great number of operating features and characteristics are exposed to a configuration tool.

When an implementation calls for only a few readers, an intelligent reader, with database memory, is all that is needed to interact with a warehouse management system. In implementations with a dozen or more readers, however, centralizing device administration provides the following benefits:

- Assignment of dynamic IP addresses.

- A management console for monitoring network traffic, and broadcasting messages to people operating print and apply stations and handhelds.

- A central library of standard filtering and event triggering schemes for download to readers.

- A control point for passing data to applications, normalizing (smoothing) the data and minimizing unnecessary data traffic to applications.

- Fast detection, swap-out and reconfiguring if a reader fails.

- An ability to upgrade firmware and configure readers remotely. These "RFID reader drivers" might vary by the type of warehouse management application.

- An ability to create and synchronize reader groups (inside and outside dock doors, etc.).

- Logging of reader and network alerts.

- A simulator for emulating tag data from readers, providing application developers with a way to configure data handling with the WMS or supply chain execution system.

Most readers will accept an Ethernet connection, RS-485 for serial device connection, or 802.11 wireless. It is possible that a network of dumb readers passing all data through a wireless network will quickly bog down, because a Wi-Fi system does not have the raw data rate to assure sub-second response time. Handheld readers equipped with Wi-Fi have utility in tag searching and inventory management, not in constant volume or high-speed applications.

Power over Ethernet (PoE) is an option with some readers. PoE is a way to supply low wattage DC power to readers by using a PoE network switch. Power consumption is then monitored and controlled along with the rest of the network elements, reducing the chance that illicit devices are on the network.

Security and Quality of Service (QoS)

Unsecured wireless networks present opportunities for eaves-dropping data. Equally vulnerable are networks of readers using IP addresses and sharing the office LAN accessible through the internet. These readers could be hijacked and reconfigured to transmit information to an unauthorized target, similar to how computers are taken over by spyware.

One solution is to have authentication a requirement before information is passed from a reader, and encryption routines in place. One method of securing reader transmissions is "silent treewalking," where EPC codes are broadcast only once, then indirectly referenced between the reader and host system. Only the host and reader know the reference system in use at the time. In a similar way, a Gen 2 reader uses a random number sent by the tag in its initial response to mask subsequent requests and receipts of the EPC on the tag.

Quality of service requirements for an RFID network exceed that of a typical office LAN. When production events depend on tag read triggers, a faulty network transmission can create production halts and downtime, seriously impacting a companies ability to meet commitments and make money.

DCs and manufacturing environments can be harsh. They require ruggedized devices and electrical components that meet guidelines for industrial safety, electrostatic and electrical noise suppression (EMI), dirt and dust, moisture protection, wash-down capability, impact resistance, etc. A network design that reduces the amount of exposed wiring by using Power over Ethernet (PoE) may be beneficial. Networks in harsh environments usually have higher error rates. Network quality can be improved through proper shielding of cables, ruggedized connectors, and enclosures for powered devices.

GLOBAL EXCHANGE OF SUPPLY CHAIN DATA

Electronic product codes are the lowest-level link in a multi-tier information model conceived by the Auto-ID Center (now EPCglobal). EPCs are assigned, catalogued, tracked and managed within a collaborative system associated with the Internet. Using an EPC, a host computer can look up via the Internet stored information about a specific item, including the manufacturer, product classification, handling, use and status in the supply chain.

EPCglobal, in partnership with VeriSign, the company that manages the Internet Domain Name Service (DNS), has begun laying the groundwork for what is called the EPC Information Network. It will leverage the Internet to support the cataloging of EPCs for supply chain partners. The overall concept and system architecture includes these other system components:

Object Name Service (ONS) – Like the Internet Domain Name Service (DNS), which provides lookups for locations on the World Wide Web, ONS serves as a registry for distributed EPC Information Services databases. Managed by VeriSign, ONS will link an EPC with the IP address of a database that stores relevant information.

Information Services – EPC Information Services are the actual data repositories used to store unique item data. These are distributed databases maintained by companies, and referenced through ONS, like how DNS points to web sites on the Internet.

EPC Discovery Service – The directory service stores EPC history. It serves as a chain of custody service, providing tracing information as a product moves through the supply chain.

Physical Markup Language (PML) – PML provides a common reference language for electronic communications amongst trading partners. Like HTML, the web page description language, and eXtensible Markup Language (XML), which is a generalized data description language, PML will extend EPC to include associated information of value to the supply chain. PML pages for each EPC can be set up, maintained and shared by the EPC product manufacturer or "owner." PML descriptors might include the following kinds of information:

- Expiry dates and safe handling instructions.
- Ingredients and composition.
- Physical properties, including telemetry information (where it is located at any moment in time).
- Procedural information, such as which processing, packaging and quality control steps have occurred.

Figure 10.5 illustrates the data flow. ALE compliant middleware intercepts the massive amount of tag data available from local readers

FIGURE 10.5

EPC information flow.

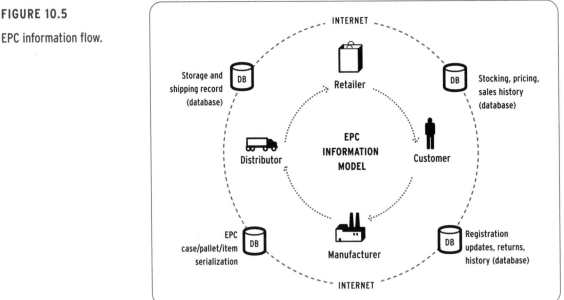

and rationalizes it into information for use with Warehouse Management Systems (WMS), Enterprise Resource Planning (ERP) systems and elsewhere. Without such a system, networks and computers would collapse from the RFID traffic generated by tags going through the manufacturing and packaging process.

EPC data architecture not only allows for faster identification of product, but supports the movement of information through the value chain and product life cycle. Ultimately this will lead to business efficiencies. An associated standardization effort, called Global Data Synchronization (GDS), is moving toward the establishment of harmonized item data.

GLOBAL DATA SYNCHRONIZATION

Global data synchronization, or GDS, is an industry-wide initiative to create standardized forms for product information that can be shared electronically by manufacturers, distributors and retailers. UCCnet facilitates the initiative. UCCnet provides services for synchronizing product data and achieving compliance. In the retail industry, mandates from Wal-Mart, Lowes and others are driving suppliers to register with UCCnet.

Global data synchronization involves a number of internal and external activities (see Figure 10.6):

→ SEE FIG. 10.6

Data cleansing – For items that it produces and trades, a supplier must adopt standard names, descriptions and identities. This activity can take the form of an internal conversion of master data files, or an application that manages the correlation of internal data files and UCCnet compliant formats. Data cleansing can be

FIGURE 10.6

GDS process.

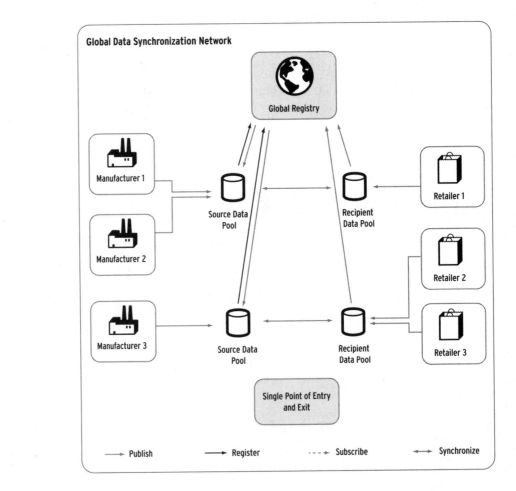

a considerable effort for a company, especially when their internal systems (accounting, manufacturing, distribution, etc.) do not represent item data consistently.

UCCnet registration and synchronization – After data cleansing, a company registers its trade items with UCCnet. This neutral data repository gives trading partners access to the item data, including product numbers and pricing. Automatic updates are used to ensure that the UCCnet registry reflects current item data for the company.

The elimination of error and the gaining of transaction efficiencies are the end goals of global data synchronization. One study reports that 60 percent of all invoices generated in the consumer goods industry have errors, with an estimated cost to correct them at $40 to $400 per incident. The monetary benefits to broad industry adoption of GDS could translate to $1 million saved for each $1 billion in sales, according to a study by AT Kearney and Kurt Salmon Associates.

Although a number of implementation issues related to GDS are yet to be resolved, the basic framework is in place for North America, as are the industry mandates for its adoption. Third party solutions and services for data cleansing and UCCnet compliance work are available.

Adoption of EPC and GDS go hand in hand. They are foundational capabilities, as shown in Figure 10.7. While RFID technology can

⊘ SEE FIG. 10.7

provide more detailed and faster tracking of product in the supply chain, it is not useful unless the exchange of data is correct. A data cleansing activity of some form is required when a company adopts EPC, just as its necessary to global data synchronization. Otherwise, a company is at the risk of simply automating a "garbage in, garbage out" system.

FIGURE 10.7

EPC and GDS are foundation activities for supply chains.

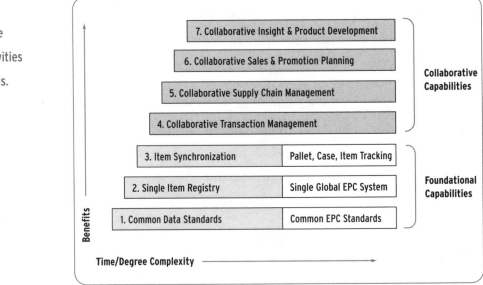

Sources and Further Information

RFID Primer, white paper, Alien Technology, 2002.

Reader Protocol 1.0, Working Draft Specification, EPCglobal, 2003.

Smart Boxes, RFID Can Improve Efficiency, Visibility and Security in the Global Supply Chain, white paper, A.T. Kearney Inc., 2005, available: www.atkearney.com.

Planning for Proliferation: The Impact of RFID on the Network, IDC White Paper by Duncan Brown and Evelien Wiggers, IDC, 2005, available: www.cisco.com.

RFID in the Enterprise, by Tom Polizzi, WCCN Publishing, Inc., 2004.

"ALE: A New Standard for Data Access, by Ken Traub," *RFID Journal,* April 18, 2005.

"RFID's Security Challenge," by George V. Hulme and Thomas Claburn, *Information Week,* November 15, 2004.

The EPC Network: Enhancing the Supply Chain, white paper, VeriSign, 2004.

Connect the Dots, white paper, A.T. Kearney and Kurt Salmon Associates, February 2004.

CHAPTER 11

Using RFID Data in the Supply Chain

By Mark Eischens,
HighJump Software, a 3M Company

COMPLIANCE MANAGEMENT 279

Tracking Compliance Data 282

WAREHOUSE MANAGEMENT 288

RFID Data Used in Warehouse Operations 291

Yard Management 294

MANUFACTURING EXECUTION 297

Data Collection Integrated to ERP 298

OTHER SUPPLY CHAIN AREAS 299

Transportation Management 299

Customer Service 300

Management Dashboard 301

Event Management 302

Supplier Enablement 302

IN THIS CHAPTER:
How RFID supports an integrated supply chain solution.

Whether introducing into a manual, bar code-enabled, or even voice technology-enabled environment, RFID will bring about a vast amount of business process and infrastructure change. Because of this significant change, implementing RFID in a 'big bang' approach (ie. implementing RFID everywhere all at once) is simply not feasible or economically viable for most manufacturers, warehouses, and distributors. A limited and highly-in-demand RFID talent pool, combined with evolving RFID standards and technologies are just a few of the factors which make this approach difficult and risky.

In contrast to the 'big bang' approach, a phased or segmented approach to RFID has proven to be the most beneficial and effective method by which companies can begin to understand and utilize RFID. In this approach, a company first identifies all of the different areas (business processes) in which RFID technology may someday benefit their operations. Recognizing that all areas cannot be met simultaneously, each is prioritized based on regulatory or customer requirements, ROI (payback), and feasibility. RFID is then deployed in serial, sequential order, based on the prioritization.

For those companies that are faced with looming RFID customer or regulatory compliance dates, the sequence in which they deploy RFID is quite simple.... compliance management must come first.

COMPLIANCE MANAGEMENT

A monumental transformation of the supply chain started taking form in 2003, spurring the rapid adoption of RFID throughout some of the largest and most well-known consumer goods companies in the world. This RFID compliance labeling 'initiative', described in Chapter 5, was monumental to the supply chain in several ways.

First, the U.S.-based initiative was led by a company who, in addition to being the world's largest consumer goods retailer, has a long and successful track record of squeezing costs out of the supply chain. This company, Wal-Mart, took the lead to introduce RFID into the realm of the consumer goods supply chain, based on their understanding of what benefits RFID could bring to themselves and their suppliers. RFID was (and still remains to some extent) a confusing and misunderstood technology by many CPG manufacturers. Therefore, it was immensely important that a company like Wal-Mart led the charge (domestically) – a company that understood what 'the next big thing' in driving out supply chain costs would be. And when Wal-Mart spoke, suppliers listened.

Second, the use of RFID technology, combined with new data standards (ie. EPC), accommodated a level of unique identification of inventory that had never been fully achieved or widely implemented before. By using RFID technology, it was now plausible to not only track each pallet uniquely, but to also track each individual case, or even each item individually throughout the entire supply chain. This fundamental change in the granularity by which inventory can be identified and tracked has the potential of paying dividends throughout many of the handling points in a supply chain.

Third, this environment (Wal-Mart driving RFID adoption) provided a 'catalyst' by which CPG manufacturers could start to re-evaluate their own internal operations for areas in which RFID could grow efficiencies. In the absence of these initial customer-led RFID programs, performing RFID R&D would have been viewed by most CPG manufacturers and distributors as risky, superfluous and misaligned with customer needs. However, this new environment of customer-driven RFID compliance not only gave CPG manufacturers the green light to start investigating RFID, but also provided them with a 'roadmap' of which technologies (frequency, tag type, etc.) were to be used.

RFID compliance management systems are designed to do exactly what their name implies – to manage RFID data and execute RFID-based, compliant labeling and verification of finished goods bound for retail customers. These systems are often referred to as slap & ship solutions. This term is derived from the most common usage of this system – whereby the RFID label is 'slapped' onto the inventory item and shortly thereafter 'shipped' off to the customer. While this process, at least at face value, provides very little ROI to the company which is applying the RFID tag, it does accomplish the main objective of physically labeling inventory with RFID tags. Additionally, slap & ship solutions address the following needs of suppliers faced with compliance requirements:

• Rapid implementation
• Minimal investment
• Initial RFID exposure (starting point for companies to become familiar with RFID technology)

RFID compliance management systems are typically small in size initially and consist of the following basic components:

• Tagging station area
• Compliance management software
• RFID label printer/encoder for both case and pallet labels
• RFID reader(s) to verify both case labels and pallet labels

Compliance management software can be implemented as a 'stand alone' workstation or 'integrated' with existing fulfillment systems (ie. ERP, WMS). Stand-alone systems are typically the least costly systems to implement, however, they offer substantially less benefits when compared to fully integrated compliance management systems.

Tracking Compliance Data

New data structures for identifying unique objects throughout the supply chain was created with the formation of EPCglobal. These data structures are described within EPCglobal's EPC Tag Data Standards definition. For example, the most simple EPC data schema, the Serialized GTIN or SGTIN, is used to identify homogenous (same-type) objects within the supply chain. At a high level, the SGTIN is essentially comprised of three primary data fields. These three fields correlate to the vendor, the SKU, and the unique serial number for that particular unit of inventory. By now, you have probably come to the conclusion that the SGTIN EPC data format is very much like a UPC with a serial number. This is a very simple, yet powerful, mechanism by which all inventory units in the supply chain can be distinguished from one another.

While slap & ship operations appear to be simplistic in nature, there exists an opportunity to generate and gather a great amount of data that can have value from the very first day you turn the system on. When properly designed and implemented, these systems can become the cornerstone by which RFID inventory tracking and distribution performance monitoring is achieved.

The point at which the RFID tag or label is commissioned and applied to the inventory unit is of paramount importance. Today, compliance labels (especially RFID compliance labels) are typically commissioned and applied only after the destination of the inventory unit is known (ie. based on sales order). Performing tag commissioning in this sequence allows the RFID tags to be designated to an exact customer order immediately.

From a sales order perspective, RFID data capture provides the following information for outbound deliveries:

- The total number of cases of each product that have been picked and labeled
- The total number of pallets (physical or virtual) that will be shipped
- The total actual number of cases on each pallet
- The makeup/composition of SKUs on each pallet
- The order in which cases were added onto a pallet (ie. layers)
- The unique identity of each case that will be sent to the customer
- The unique identity of each pallet that will be sent to the customer
- The date and time in which the labeling process started and ended for each case and pallet
- The date and time in which each of the cases and pallets were fully verified for RFID readability
- The date and time in which each of the cases and pallets were loaded onto the truck

- The sequence in which pallets were loaded onto the truck
- The specific dock door or trailer that the order was loaded at or on

Using many of the data points captured above, it is quite feasible to use compliance management systems to achieve greater operational efficiencies and shipping accuracy. Below are just a few of the questions that can now be answered with RFID data that may not have been possible before.

1. How often are shipments leaving the facility without correct compliance labeling?

2. How often are operators moving pallets into the wrong truck?
   - How often are orders picked correctly but loaded incorrectly?
   - Where and how did the mis-match of inventory occur?
   - What movements of inventory led up to the issue?

3. How many times were pallets moved onto a truck and later moved off?

4. How often are orders picked inaccurately?
   - Shorts?
   - Overages?
   - Incorrect SKU?
   - How well is FIFO rotation occurring during pick operations?

5. How well are mixed-SKU pallets being built (ie. are heavy SKUs being put on bottom and lighter SKUs on top to avoid damage)?

6. Are optimal pallet configurations being achieved? Are pallets being built properly, based on standard operating procedures or best-practices?
   - How well are material handlers consolidating pallets?

7. How well are my RFID printers and tags performing?
   - How many reprints (reprinting labels) were required?
   - Are certain SKUs proving more difficult to tag and/or verify than others?
   - If so, which SKUs are they?

8. What type of read rate are each of my items receiving (cases and pallets)?
   - Are the current RFID tag placement locations providing adequate performance?

The questions above are indicative of what information you have available before the order is even shipped to your customer. A properly designed and implemented compliance management system will provide you with this information. Additionally, there is a wealth of information that becomes available once the inventory units traverse the supply chain.

For most, if not all, compliance management systems today the point at which RFID tags are generated (commissioned) and applied to inventory occurs during or shortly after the customer order has been released for picking at the warehouse or distribution center.

While labeling so far back within the warehouse operations limits the amount of initial payback from RFID tags, it does provide an ideal environment, from a data perspective, by which RFID tags can be generated. For starters, it is at this time that the ultimate destination (customer and ship-to location) is known for that particular inventory unit. Today, this is very important as it is quite likely that your customers have different labeling requirements (human readable, bar code, and RFID). Hence, the format, size, text, and technology (bar code, RFID, or both) of the label you apply will depend on the customer. Another benefit to printing at this time is the ability to directly associate the RFID labeled inventory unit to an exact customer order. Generating this logical association at the time you print RFID labels ensures that you will be able to verify order accuracy during your pick, pack, and ship processes.

Additionally, at this time, the final packout has been executed for each inventory unit that will be RFID labeled and the inventory unit will not undergo any additional packaging transformations (re-pack, consolidation, over-pack) before it is shipped.

For most companies today, compliance management is the first, small incremental step towards utilizing RFID in the supply chain. Certainly it is important to successfully implement and integrate RFID compliance management for the sake of your customers. But beyond that, it is important because it represents the first time in which your company has delved into RFID technology. It provides a vitally important learning curve to you and your company – a learning curve that will help guide you through future adoption and integration of RFID technology into other areas of your business.

Compliance management is important because it is the mechanism by which most, if not all, EPC RFID data is 'front loaded' today. In other words, there has to be a starting point by which supply chain operations is monitored. Compliance management is the most plausible and well-positioned mechanism to do just that.

And finally, compliance management can provide immediate benefits to you and your company. For example, EPC RFID technology delivers a level of automated inventory identification that can be used to improve order shipment accuracies. Whereby common non-RFID systems rely on pallet-level bar codes and visual verification of pallet contents (cases), RFID EPC systems can audit each and every case individually, pallet by pallet. This increased level of identification, combined with automated verification, can enhance shipping processes for both throughput and accuracy. Essentially, what you end up with is a case-level audit trail for each and every case on each truck. This audit trail will come in handy later on in instances where order discrepancies (shortages, deductions, charge-backs) arise within your distribution network.

As you can see, labeling your inventory for RFID compliance management is just the beginning step towards utilizing RFID in the supply chain to drive operational efficiencies and effective inventory management.

WAREHOUSE MANAGEMENT

The challenges of managing distribution include ever-increasing customer expectations, compliance requirements, growing competitive pressures and evolving business models. Meeting customer service demands sometimes runs counter to the need to reduce operational expenses. Warehouse Management Systems (WMS) provide that balance with technology that harnesses information and uses it for intelligent work direction.

For decades, warehouse management systems have driven out distribution costs by optimizing the flow of goods, maximizing resource utilization, and increasing inventory accuracy. To achieve these new levels of performance and intelligence, these systems are fueled by massive amounts of data. This data comes from a variety of business systems such as ERPs as well as input from human operators and automation equipment.

So what ramification can RFID data have on a warehouse? First, we will examine the effect of RFID data within the warehouse itself.

RFID, in concert with EPC standards, promises to provide unique identities of inventory units (case, pallet, item, etc.) throughout the supply chain, including the warehouse. This concept is quite radical as compared to the way in which many warehouses currently operate. Today, many organizations manage the majority of the products at the SKU level, whereby a quantity of a SKU is known for each location in which the product is stored. For example, "there are 6 cases of item A100 in location H690." While this method satisfies many of the requirements needed to execute current customer delivery of the product, there are several inherent deficiencies that may become increasingly important to address via RFID in the near future for both manufacturers and distributors alike.

First, the lack of unique identification on each package (carton) ultimately means that each carton of that SKU must be viewed by the WMS as 'equal' to the other cartons of that same SKU. Therefore, planned activities such as picking must be performed without the knowledge of specific carton-level attributes such as manufactured date, manufacturing origin, engineering revision, batch/lot, or even warehouse receipt date. As an example, instituting and enforcing inventory rules such as First-In-First-Out (FIFO) is difficult, if not impossible, in these environments. Additionally, in the absence of lot numbers or serial numbers, products that flow through the warehouse cannot be tracked effectively back through distribution to manufacturing for important information that may be used to recall product, identify shrinkage points, or even prevent spoilage.

By pushing RFID data from manufacturing to distribution, the WMS may now plan and execute activities based on actual, discrete, granular inventory information. FIFO management of inventory becomes more feasible. Tracking product flow in and out of the warehouse with RFID creates far greater capabilities to identify a small subset of products, their location, and even direction of flow – thereby aiding recall campaigns, real-time notification of shrinkage, and preventing spoilage issues associated with excessive inventory dwell times.

It is important to note that some warehouses currently do identify and track inventory at greater levels of granularity than only the SKU. For example, high-tech products are commonly tracked with serial numbers while lot numbers are typical of food and beverage products. These industries have justified the additional labor costs necessary to generate and track these items. These same labor costs have historically prevented many distribution companies from instituting serialized control of inventory as this increased level of identification, while beneficial, simply did not outweigh the costs.

To understand whether RFID can help reduce these costs and provide additional value, let's next examine the effects RFID and RFID data has on a few of the operations within a warehouse.

RFID Data Used in Warehouse Operations

Many operations within a typical warehouse rely on real-time input from human operators. These operators ultimately respond to instructions sent down by the WMS and utilize proven technologies such as manual input, bar codes, and even voice to verify data such as SKUs, quantities, or inventory locations. While SKUs and locations can be collected accurately via bar codes, quantities are most often determined by simple visual verification by the operator. This operator-dependent scenario is manifested throughout the warehouse – from order receipts to cycle counting to picking.

The introduction of RFID data within the warehouse poses several opportunities to reduce the reliance on operators for accurate inventory control. Several of these opportunities are listed below.

Receiving – Data from RFID tags on each case and pallet is captured and automatically reconciled against expected receipt quantities provided by supplier. Shrinkage, item substitutions, quantity discrepancies, and shipping errors are identified at point of receipt and recorded against supplier scorecards.

Putaways – RFID data captured during the putaway process ensures that the entire quantity is accounted for during the putaway process. Rather than relying on visual counts of cases and scanning of putaway locations, RFID tag data read at each storage location guarantees that the actual putaway location for each case is known at all times – this is especially important information to have if the actual location differs from the planned putaway location.

Movements and transfers – Capturing detailed inventory movement information is the first step in understanding traffic patterns, operator efficiencies, and equipment utilization. Imagine every inventory item reporting it's movement throughout the entire warehouse by generating a time-stamped audit trail as it passes through each area. Analyzing this information provides a completely new view of how efficiently goods are handled, operator trends that are occurring but have previously been 'invisible', immediate recognition of shrinkage conditions, and also opens up new opportunities for identifying when best-practices are violated or anomalies occur during planned and un-planned movements.

Kitting – RFID data can be used to increase accuracies and reduce shrinkages for kitting operations by automatically verifying actual contents against planned during the kitting operation. Identifying shrinkage, incorrect components, and overages/underages real-time leads to better inventory control and quality control. Additionally, carrying forward RFID data from component goods into finished assemblies creates enhanced capabilities for recall management and returns processing.

Picking – Leveraging the data within an RFID tag, specific individual inventory items may be quickly located by using an RFID handheld reader in a similar fashion to a Geiger-counter. Audible tones indicate to the operator where an inventory item is amongst numerous similar items within a certain location.

Shipping – Improved accuracies in shipping operations is a direct result of properly harnessing RFID data to ensure the right quantity of the right product, is loaded on the right truck. RFID data captured at point of shipment provides a solid audit trail which can later be leveraged should discrepancies occur at your customer's receiving door.

Returns – With an idealistic RFID data audit trail reaching back to the manufacturing process, returns can be managed more effectively and efficiently within a warehouse. Returned or damaged inventory can be safely isolated from other finished goods inventory, recorded and analyzed against the logistics entities who handled it (ie. identify point in which damage occurred), and can even be routed back to their (manufacturing) origin.

RFID data which is made available to warehousing operations can have a dramatic impact on distribution efficiencies, performance, and accuracy. Conversely, the actions warehouse managers take within their four walls may ultimately have a direct impact on their supply chain partners, regardless of whether those partners are located upstream or downstream from the warehouse. Improved flow-through, greater levels of quality, accurate procurement of goods, and accurate delivery of goods all have rippling effects that eventually are felt by trading partners.

Furthermore, sharing RFID data with these trading partners allows them to gain a bigger picture view of inventory movement – this information helps them understand better the origins of the inventory they handle, the distribution channel in which their products are delivered, as well as the consumption trends (ie. how fast is the product moving off retail shelves?)

Improving communication amongst these partners, and leveraging the RFID data tied to each inventory unit, is instrumental in driving cost and inefficiencies out of the supply chain.

Yard Management

Its easy to see why companies want to extend RFID outside their distribution center walls. Yard Management, the tracking, management, and optimization of mobile inventory container (ie. trailers) is an area in which RFID data can be utilized very readily. Already, yard management software, integrated with a WMS, can optimize the staging of shipments and transport. For example, a yard management system can instruct drivers to park nearby the north side of a building if a north side dock is going to be used. A pick logic can be employed for retrieving trailers from a yard. A trailer with the earliest arrival data, for example, would be scheduled first to the dock.

RFID dock portals provide vision for the yard management software system. Additionally, RFID tags used on trailers in the yard can be read by yard workers equipped with handheld readers. Alternatively, Real Time Locating System (RTLS) RFID technology may be utilized whereby the location of trucks, equipment, and even personnel can be tracked instantaneously anywhere in the yard. Similar to GPS, but on a much smaller scale, RTLS provides additional automation and exception management opportunities.

One interesting aspect of utilizing RFID within a trailer yard is that the yard itself can become more flexible and allow for more 'errors' without incurring decreases in accuracy or efficiencies. For example: a trailer enters the yard and the truck driver is directed to put this trailer in location B11. Upon arriving at location B11, the truck driver notices that another trailer is already there. The truck driver then locates the nearest empty location, B15, and drops the trailer off. Without RFID technology, and without great incentive for the truck driver, this new location would often times not be reported back to the yard manager. Hence, when a yard jockey was dispatched the next day to retrieve this trailer, they would spend precious time trying to locate and positively identify the trailer. This time spent is a result of the 'error' made earlier the following day. This 'error' was really only a discrepancy between planned activity and actual activity and certainly wasn't intentional or malicious. Nevertheless, it ended up causing confusion and wasting time and ultimately took human operator intervention to identify and correct the issue.

Integrating RFID technology to the same scenario eliminates many, if not all, of the issues described above. For example: If all trailers in the yard were RFID-tagged, then the truck driver with the new trailer would never have been dispatched to location B11 – the real-time view of yard trailer locations would have alerted the yard manager that this location was occupied and the truck driver would have been directed to another location. And, for sake of argument, imagine if the trailer currently in location B11 did not have an RFID tag and the truck driver was again directed to put the new trailer in location B11. Upon arriving, the truck driver again drops the trailer off in location B15. This time, however, the RFID system captured this drop real-time, updating the yard manager of the situation. When the yard jockey is instructed to retrieve the new trailer the next day, he will be assured that the location, B15, that he is dispatched to has the correct trailer waiting for him.

Yard management software, combined with RFID, creates a powerful, yet flexible way to optimize yard space, maximize operator/jockey performance, and execute on-time deliveries to the receiving dock of the warehouse.

MANUFACTURING EXECUTION

Efficiently managing production operations is a difficult task in today's fast moving consumer goods environments. Market pressures require maximum flexibility. Financial pressures require continual process improvement. Accurate information and MES software can help guide manufacturers to make good decisions. Features of such a system include:

Shop floor data acquisition and execution – By using smart labels and intelligent shop floor terminals, the MES can instrument work in process, scrap and production reporting, work orders, drawings and specifications, lot and serial number tracking and quality control.

Management visibility and event notification – The MES provides alerts and notifications, wellness views, key performance indicators, production data in real-time, all in configurable dashboards and reports delivered on-line or via mobile handhelds.

Device integration – By integrating RFID data with portable data terminals, mobile phones, pagers, SCADA (supervisory control and data acquisition) information from scales, PLC networks, conveyors, sorters, palletizers and gages, the MES can provide a consolidation point for management and useful information for product tracking and traceability.

An MES leverages enterprise resource planning (ERP) systems by providing intuitive execution capabilities based on a real-time, lean execution philosophy. MES data is then fed back to the ERP for business planning.

Data Collection Integrated to ERP

Many ERP systems can accept shop floor data in normalized form. A data collection software layer provides this by providing a common interface for managing bar code, RFID, SCADA, measurement device and Web-based data. Data collection software can be used to integrate data involved in these operations:

- Labor reporting
- Inventory tracking
- Directed picking
- Lot control
- Maintenance and repair operations
- Time and attendance

• Shipping verification
• Electronic Kanban

OTHER SUPPLY CHAIN AREAS

In addition to warehousing and manufacturing, other areas within the supply chain may leverage RFID data for increased levels of performance. They include:

Transportation Management

Focused on route analysis and optimization, a transportation management system is one component of a supply chain execution system that can effectively leverage EPC data. The goal is to facilitate the speedy and accurate movement of goods, and link up with the EPC to add track and trace capability. Transportation management software provides web-based tendering, tracking and supply chain visibility. Shipping specialists and carriers have accurate information for proof of delivery and fast resolution of problems.

RFID data can provide enhancements to these areas of functionality:

Order consolidation – Split shipments or load consolidation does not cause delays from having to manually locate and record what product went where.

Load optimization – Automated by the system, loads are verified at the dock portal.

Cargo Monitoring – Automated monitoring of inventory conditions can alert suppliers and carriers of impending spoilage due to temperatures, pressures, humidity, etc.

Tracking – Dock portal reads transmitted through the system supports tracking by item and time stamp.

Customer Service

Most customer concerns tend to revolve around order status issues, including order receipt, time to delivery, and shipping information. By having customer service software linked with your RFID-enabled supply chain execution system, companies can be proactive with their service. Web-based, self-service portals can be used to provide customers with all the information, linked to the EPC. According to Forrester Research, nearly 60 percent of Web-based customers or partners monitor the status of orders during processing and shipping.

By putting customer service information on the web, companies can provide:

- Updates on inventory levels
- Exact date and time that orders were received, picked, packed and shipped

- Package tracking numbers
- Shipping, delivery and logistics data
- Round-the-clock answers to the most common questions

Management Dashboard

Aggregating EPC data with other information in the supply chain, dashboard software provides operations managers with key performance indicators, plus a context for fact-based real-time decision making. Web-based dashboards can be tailored to the role and responsibilities of each manager. They can contain rollups of inventory levels, order volumes and order status, with drill downs, comparative metrics, and alerts based on operational priorities.

Dashboard analytics leverage the power of real-time operational information and the ability of RFID to provide it. Managers can choose their views and use advanced analytics to close the information gap between functional areas and enterprise goals. By providing quick and clear comparisons, dashboards facilitate insight into distribution dynamics. For example, comparing order volumes on a week-by-week or day-by-day basis can significantly improve stocking levels and procurement pricing of consumables such as packaging materials. Likewise, comparing cycle times across various warehouses can help to optimize asset utilization across the enterprise.

Event Management

Events in one part of a supply chain can have a ripple effect. Slowed by manual processes and insufficient visibility into affected areas, remote events can have direct and significant consequences. Event management software provides notification mechanisms, such as email, pager, fax and phone, integrated with workflows to facilitate rules-based response to supply chain events. This includes escalation up the management chain or to supply chain partners.

Event detection capabilities can help identify and reduce stock-outs, backorders and chargeback situations. By orchestrating prompt and effective response mechanisms, customer service is enhanced and customer relationships optimized. Event rules can be easily set up. For example, if a carrier is scheduled to pick up a load and is more than two hours late, the system can automatically pass the load to a backup carrier. Alerts can be used to notify when a shipment arrives ahead of schedule, and set next steps in motion. By integrating events across the supply chain, including triggers based on RFID read verifications, event management software can facilitate efficient response and minimal disruption to the flow of goods.

Supplier Enablement

As the speed and complexity of order-fulfillment operations escalates, a company's ability to execute to meet customer demand becomes directly determined by its suppliers. A supplier management software system, leveraging RFID data, can facilitate the supplier collaboration, especially with a global supplier network. Elements of the supplier management system include:

• Intelligent replenishment management to reduce inventory levels, shrinkage and obsolescence.

• Speedy reconciliation of purchase orders, with workflows associated with PO's being rejected, not responded to, or late response, offering choices of other items for replenishment, short shipping or alternate suppliers.

• Status indicators for shipped, unshipped and short-shipped goods.

• Web-based ASN receiving to increase facility flow-through and eliminate EDI transaction fees.

• Response to order point triggers

• Receipt acknowledgement

• Shipment tracking

Sources and Further Reference

For more information, see www.highjumpsoftware.com.

CHAPTER 12

A Partner's Approach to the Supply Chain

By Greg Gilbert
Director, RFID Solutions and Strategy, Manhattan Associates

WAREHOUSE MANAGEMENT 306

DISTRIBUTED ORDER MANAGEMENT 308

TRADING PARTNER MANAGEMENT 310

TRANSPORTATION MANAGEMENT 311

RFID IN A BOX® 313

INTEGRATED LOGISTICS SOLUTIONS™ 315

IN THIS CHAPTER:
How WMS RFID capabilities integrate with supply chain solutions.

As many enterprises are now realizing, RFID is a transformational technology capable of delivering substantial business value well beyond basic compliance. WMS companies, such as Manhattan Associates, can help organizations realize that value with their RFID-enabled supply chain solutions by integrating EPC data into a complete warehouse management system.

WAREHOUSE MANAGEMENT

The increasing complexities of warehouse operations make it critical for companies to have solutions that synchronize business processes and provide competitive advantage. Warehouse Management solutions meet these needs with a range of capabilities for labor, slotting, load management, billing and performance management. This helps an organization take on the challenges of changing customer requirements, globalization, compliance initiatives and multiple distribution channels.

The Warehouse Management solution provides real-time synchronization of information and products within the warehouse and yard. Receiving and shipping can be streamlined to facilitate cross docking and expedite back-ordered product. The solution provides traceability down to the lowest level of detail, including country of origin, lot number, serial number and date code. Cycle counting capabilities make it possible to eliminate annual shutdowns for physical counts.

EPC data gathered from RFID tags through the Warehouse Management solution is cleansed by the Integration Platform for RFID, providing the detailed data used in tracking and meeting compliance. Integration of RFID enables increased speed and efficiency of warehouse processes, including receiving, putaway, allocation, inventory management, picking, packing and shipping validation. It facilitates faster throughput at cross docking and reduces shipping errors.

The Warehouse Management solution can integrate a number of components, including:

Labor Management – The Labor Management component helps improve employee performance by balancing workload and measuring productivity. It can track all warehouse labor activities and measure them against an organization's baseline standards. This helps extend targeted training and incentive programs.

Slotting Optimization – The optimal placement of products within the warehouse is based on historical and current order demand. Slotting Optimization helps increase workforce efficiency, shorten order fulfillment time and minimize accidents. This solution can help slot new products, organize a pick layout or re-work target areas. It allows you to run scenarios that provide immediate feedback on ROI before going to implementation.

Load Management – This solution provides visibility into yard activities so you can plan and execute all loads entering or leaving the facility. It helps with yard planning, incorporating all of the complex requirements associated with shipping and receiving. Carriers and shippers can schedule appointments using EDI or a Web interface. You can also send alert notifications to proactively correct issues before they cause delays.

Billing Management – This solution allows you to track all inventory handling, storage, fulfillment and transportation activities by business unit or client. Logistics service providers, especially third- or fourth-party logistics providers, can manage customer contract requirements to improve billing accuracy and service, preview invoices, audit charges and apply a percentage discount.

DISTRIBUTED ORDER MANAGEMENT

Distributed Order Management solutions extend visibility and control across the entire supply chain. The solution routes orders which require RFID to the appropriate distribution center. When integrated with the Warehouse Management solution, Distributed Order Management helps improve order fill rates while reducing inventory and operational costs. It provides instant communications with trading partners and coordinates the entire order fulfillment lifecycle. Functions include:

- Order capture and aggregation through a Web-based portal;
- Demand prioritization to ensure important customers are given priority;
- Order sourcing from supply channels;
- Order allocation per line item, prioritized according to business rules;
- Scheduling and shipping optimization;
- Purchase order (PO) recommendation;
- Exception management and substitution based on business rules;
- Order orchestration workflow.

Distributed Order Management solutions provide a single repository in which orders can be aggregated and the lifecycle of every order can be viewed and managed in real-time. It can help streamline the process of accurately matching demand to inventory availability, including in-stock inventory, back orders, inbound shipments, supplier POs and unexpected receipts. It also benchmarks and monitors key performance indicators such as forecasting accuracy, order-to-delivery time, inventory turns, on-time delivery rates and cash-to-cash cycle rates.

TRADING PARTNER MANAGEMENT

Companies with extended networks of suppliers and customers face a daily challenge of synchronizing business processes with that of their trading partners. Trading Partner Management solutions offer a series of components for working with suppliers, logistics hubs, carriers and retail sites to increase fill rates, reduce safety stock, speed time to market and change production to better reflect demand and market trends. When integrated with RFID, the solution provides improved, near real-time visibility into supply and demand conditions and a better proof point for dispute resolution.

Electronic POs ensure that suppliers are always working against the most current documentation, reducing the potential for error. Advance ship notices (ASNs) can be generated and sent electronically, providing accurate information about inbound shipments. Manufacturers can use this solution to print compliant bar code labels, RFID tags and shipping documentation via the Internet.

Supplier Enablement – This component of a Trading Partner Management solution extends supply chain execution capabilities to your trading partners to enable synchronization of business processes.

Logistics Hub Management – This solution provides visibility into distribution hubs and 3PL activities. Hubs are able to manage inbound and outbound ASNs electronically, as well as support drop-ship programs.

Carrier Enablement – This solution provides real-time visibility into inventory in transit. Carriers can provide shipment status updates through an online browser screen.

Customer/Store Enablement – This solution enables customers and retail outlets to place, update, track and confirm orders online. It provides controlled access to product availability and order status, thereby reducing call center and customer service overhead. Orders can be placed electronically and appear the same as when placed via EDI. Upon receipt of product, customers can confirm exactly what goods are received and offer feedback regarding the quality, timeliness and service level provided. This solution also offers ways to measure fill rates, order cycle time and on-time percentages against key performance indicators (KPIs).

TRANSPORTATION MANAGEMENT

Integrated transportation and logistics management have come into sharp focus as effective strategies for suppliers, manufacturers, distributors, retailers and service providers. Transportation Management solutions span the entire spectrum of processes related to a global transportation network, including procurement, planning, execution, load management and performance measurement. Integration with EPC data provides RFID-enabled yard management, container tracking and environmental/temperature sensors.

Transportation Procurement – A Web-based solution is used to manage the bid process and contract sourcing for ocean, air, rail and surface transportation. This solution helps you distribute bid information, receive and analyze bids and award the contract entirely online.

Transportation Planning & Execution – Provides a way to formulate a plan and optimize the day-to-day efficiency of the transport network. It provides automatic alerts to help you respond to delays or changes in flows and avert potential delivery problems.

Load Management – With Hours of Service in the U.S., the Working Time Directive in Europe and other regulations, carriers today have less tolerance for costly delays. To meet this challenge, the Load Management solution can help plan, execute, track and audit incoming and outgoing shipments through efficient appointment scheduling, carrier communications and coordinated yard management.

RFID—when integrated across the supply chain and throughout a WM system—optimizes operational efficiency and maximizes visibility throughout the extended enterprise. Solutions such as

Manhattan Associates' RFID in a Box® can provide the flexibility to integrate into any business model or application environment and can result in substantial business benefits beyond basic compliance.

RFID IN A BOX®

Manhattan Associates developed RFID in a Box to help a company easily implement RFID. It provides all the components and services required to successfully deploy a targeted or enterprise-wide RFID initiative, including software, middleware, hardware, implementation and training.

The solution's Integration Platform for RFID includes three primary components—EPC Manager, Enterprise EPC Manager and Integration Manager. The EPC Manager captures and tracks unique EPC read data at all levels of an enterprise so it can be used with other solutions and existing systems. It provides EPC standards compliance through functionality that includes EPC allocation, commissioning, product and order aggregations, inbound and outbound shipment validations, support for 64- and 96-bit tags and data collection.

While EPC Manager provides local capabilities for the integration of RFID, Enterprise EPC Manager acts as a controller and data repository. Enterprise EPC Manager enables you to autonomously manage ranges of serial numbers at local facilities to ensure duplicates are never created, without requiring you to manually configure ranges at each site. The solution also maintains a standard event interface to facilitate the integration of EPC events within the enterprise and from trading partners. As a result, users are alerted and additional workflows and processes are triggered based on expected events occurring—and even not occurring.

The Integration Manager enables integration capabilities with either Manhattan Associates' solutions or other software applications. It provides flexible reader support, configurable interfaces and extensibility and allows integration of RFID data into multiple applications—regardless of platform, data format or communications protocol. The Integration Manager also provides connectivity to the physical RFID infrastructure of readers and other devices such as motion sensors, photo eyes and light trees.

These components are built with integration in mind, utilizing modern standards-based technology such as Web Services. This allows the capabilities of the Integration Platform for RFID to be leveraged by other applications and be seamlessly integrated into existing enterprise application architecture.

The services component of RFID in a Box® includes product assessment and implementation services. The product assessment service provides pre-implementation simulation testing of various tag, reader and antenna configurations to determine the best solution for various product, case and pallet requirements. Implementation services help a company align their RFID strategy with their business model and execute against specific operational goals.

INTEGRATED LOGISTICS SOLUTIONS™

RFID in a Box® is a component of Manhattan Associates' Integrated Logistics Solutions, a source-to-consumption solution that enables companies to streamline operations and transform the supply chain into a single, well-coordinated business process. See Figure 12.1.

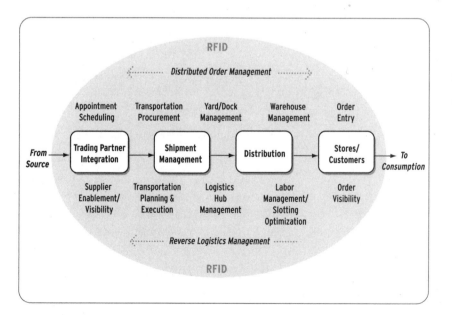

FIGURE 12.1

RFID integrated into a source to consumption logistics solution.

Developed for companies of every size and across multiple industries, Integrated Logistics Solutions can be implemented as an integrated whole for maximum value or may be deployed as individual point solutions. Integrated Logistics Solutions include Distributed Order Management, Warehouse Management, Transportation Management, Trading Partner Management, RFID in a Box®, Reverse Logistics Management and Performance Management.

Sources and Further Reference

For more information, see www.manh.com.

CHAPTER 13

Lessons Learned

FOOD INDUSTRY   **318**

Case 1: Chips in the Freezer   **318**

Case 2: Taking the Freeze out of RFID   **324**

Case 3: A New Generation Supply Chain for Snack Foods   **330**

ELECTRONICS INDUSTRY   **338**

Case 4: Smart Partnering for a High-tech Titan   **338**

MILITARY SUPPLIER   **345**

Case 5: Transportation Supplier Rides the RF Wave   **345**

IN THIS CHAPTER:
Case studies of
early adopters. What
they've learned.

There's no doubt that the fastest way to learn is by doing. Here are profiles of several actual companies embarking on the RFID journey, and the lessons they are learning.

FOOD INDUSTRY

Case 1: Chips in the Freezer: An RFID Pilot for a Frozen Food Master

For a half-century, this American household name has been synonymous with popular cakes and frozen baked desserts. Now a $15 billion-plus producer of consumer packaged goods (CPG) ranging from food and beverage to apparel and household products, the company ranks in the Fortune 200, operates in 55 countries, and markets well-known branded products in 200 nations.

Designated as one of Wal-Mart's top 100 suppliers preparing for the January 2005 RFID compliance deadline, the company initiated RFID pilots with the goal to build a technical platform that could be systematically leveraged across the entire organization. To implement the technology and its systems and processes in a coordinated, controlled approach through a central office, the company created an RFID program management team to interface with each business unit for rollout. After an analysis of the products that ship from the company's third party logistics provider to the Wal-Mart distribution centers in Texas, the team selected two divisions within the company to launch the pilot project.

RFID Challenge

Through the pilot process, the company planned to gain insight into associated costs, benefits and opportunities. Its objectives also were to study and ensure tag performance with different types of products and packaging (liquids, metals) in different environments (ambient/cold), and to eliminate manual processes wherever feasible. Over a six-month period, the company had tested different RFID printers and experienced mixed results. Familiar with other Printronix products, the company contacted Printronix for a demonstration of an RFID desktop printer, which, after review and testing, was selected to assume RFID label printing for the pilot project. Partnering with its third party logistics provider and supply chain execution and optimization specialists Manhattan Associates, the company developed an efficient RFID pilot process flow and deployed the Printronix printers.

Solution

Using Manhattan Associates' RFID in a Box®, a comprehensive suite of RFID-enabled and EPC-aware supply chain execution and visibility solutions coupled with a powerful and flexible RFID data management and integration solution, the company had an off-the-shelf solution for simplified implementation of RFID.

The pilot was developed as a de-palletization/re-palletization process, known in the industry as "slap and ship," beginning at the logistics partner warehouse. This involved setting aside mixed pallets at the end of the normal shipping process, and adding a final RFID application step. The pilot analyzed how tags should be applied, the average time involved and efficiency of the process, and studied SKUs with RF-unfriendly contents such as liquid and metals. After analysis, the company re-engineered various steps in the labeling process to improve efficiencies where needed.

Optimal RFID label placement was established and 64-bit Alien Class 1 squiggle RFID tags were selected for the 4" x 2" case label. To avoid the inefficiencies of manually applying large batches of labels to cases, the "slap and ship" process was automated using the Printronix SLPA7000 RFID encode- print-and-apply RFID printer applicator. The company continues to use the Printronix SL5000e MP™ desktop RFID printer for less intense pallet labeling. Highly compatible with Manhattan Associates' RFID in a Box® solution, the SLPA7000 was co-developed with veteran applicator provider FoxIV Technologies to meet the requirements of large compliance CPGs seeking high-volume RFID automation. Both the SLPA7000 and the SL5000e MP include multi-protocol UHF encoders to encode multiple tag classes and sizes. Used together, these UHF products would address the needs of the company to tag high-volume, high-speed cases, then individual pallets for shipment to select distribution centers. Automated label application was determined most efficient and cost effective.

Operation

Pallets ready for shipping to the designated Wal-Mart DCs are loaded next to the RFID conveyor. The selected pallets are broken down and placed onto the conveyor. This process typically consists of 40 or more cases per SKU – and occasionally with multiple SKUs per pallet.

Step two is a multi-step process. See Figure 13.1. As each SKU or case runs on the conveyor, adjustments made to the SLPA at the start of the run accommodate case size and label orientation. Each case size requires a different set-up for the SLPA, and a different set of EPC numbers for the WMS to index to that SKU. As cases are loaded onto the conveyor, they move toward the SLPA7000 printer applicator, which labels the cases with an 64-bit Alien Class 1, 4" x 2" RFID label.

→ SEE FIG. 13.1

EPC numbers cross-referenced in a look-up table containing specific SKU numbers are allocated to case labels as products run in batches on the conveyor. As each product/SKU is placed on the conveyor, the matching EPC number is delivered to the printer,

FIGURE 13.1

Frozen food company de-palletizing, labeling and re-palletizing process.

which encodes the RFID tag and prints the relevant information with a bar code on the label. A conveyor sensor detects a passing case and sends a signal to the SLPA to apply the RFID label to the case passing by. Using another box conveyor sensor to verify that a tag is on the case, cases are run through an RFID reader portal, and

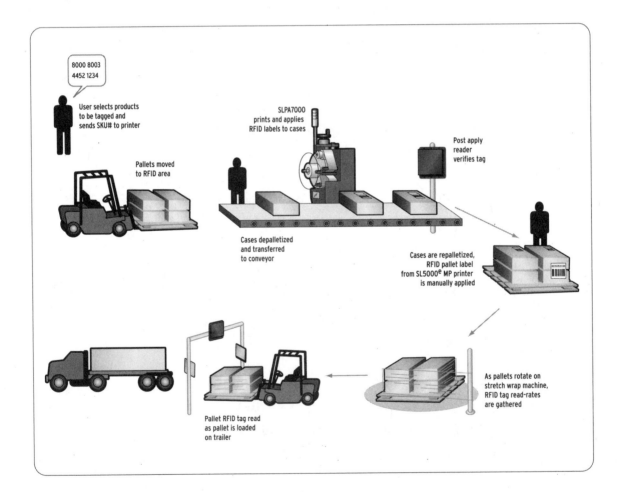

verified. In case product is not labeled because of multiple consecutive bad tags on the label roll (the label is voided and not applied), the case is rejected and the conveyor stopped. An operator removes the package and places it at the beginning of the conveyor line to be reprocessed by the SLPA.

In the next phase of implementation, this rejection step will be automated and the reject cases will be re-routed back into the RFID conveyor line, where the SLPA will then apply the tag.

Finally, cases are re-palletized, stretch wrapped, and labeled manually with a SL5000^e MP desktop printer printing a 4" x 6" Class 1, RFID pallet tag. A final post-process read of the pallet tag is made for verification, and the pallet is ready to ship to the designated Wal-Mart DC for compliance. As the pallet is read, an ASN will be sent to Wal-Mart for entry into their system, to verify against the actual shipment when it arrives from the company.

Conclusions

As RFID progresses, the company foresees increased supply chain visibility, tracking individual cases as they move from manufacturing, to distribution, to stores. Data flowing back to the management team at key points in time are near real-time, and planners will be able to use the information to better plan and forecast, allowing the company to shrink safety stock.

The company has decided on an incremental approach to adopting the technology as standards and capabilities continue to quickly evolve. It will focus the enterprise RFID team on common solutions: middleware, label/tag application process, equipment selection, purchasing leverage (equipment, tags). It also plans to continually improve the tagging process. This may involve tagging products in-line, or upstream of its current compliance model of application. As this technology matures, optimal label applications will evolve; applying RFID tags as part of the manufacturing process and increased automation will improve the company's ROI in RFID.

According to the company, the next step is to scale the process to introduce the next wave of products requiring tags for 2006. Using its experience from 2004, the company is now positioned to move its RFID implementation forward, cost effectively, and efficiently.

Case 2: Taking the Freeze out of RFID

As a large manufacturer of both frozen and refrigerated foods based in the U.S., this company makes thousands of different products for its customers including retailers like Wal-Mart. With its high quality standards and efficient manufacturing processes, the food manufacturer provides customers with premium taste products at a desirable price. Taking charge of its destiny, the company tackled the Wal-Mart RFID mandate with the decision to implement the technology using some existing management systems.

The food manufacturer created a process-driven technique using its existing host system network to interface with its enterprise resource planning (ERP) system, eliminating the need for a third-party integrator. Having extensive experience with an automated bar code manufacturing process using non-RFID print and apply systems, the IT staff devised how the application would replace the current systems with RFID-enabled printer applicators for fast production runs.

RFID Challenge

To start, the company selected a stretch wrapped, twin-pack frozen product to fulfill the immediate Wal-Mart RFID requirements – production lines that represented the most challenging smart label application because of the twin-pack's shape and irregular surface. If the smart label could function while accommodating the twin-pack's curved surface, the RFID-enabled printer applicators could be implemented plant-wide. Also, with the fast production rates, the RFID printer applicator would be subjected to the toughest product throughput surges resulting from an operator upstream shutting down to replenish the stretch wrapper's supply.

Using the existing host network system and their device management software, all the RFID devices were connected with Ethernet. Without a third-party integrator, the group set out to find an RFID printing solution that could encode multiple protocols (EPC Class 0, 0+, and 1) and bring proven reliability and upgradeability for asset management. Some systems that had been considered offered limited and/or unacceptable solutions such as applying defective smart labels to the product or using a mechanical reject arm, which even allowed the smart label to fall to the floor, causing a potential safety hazard.

Solution

The company turned to Printronix for an Ethernet-enabled Smart Label Printer Applicator. The Printronix SLPA7000 is the industry's first RFID-enabled, fully integrated encode, print and apply system and incorporates the first integrated tag management system that allows bad and quiet smart labels to remain on the liner. The company also selected the 64-bit Alien Class 1 squiggle tag using the 4" x 2" label format to encode the EPC data and print a bar code with human readable text. With this solution, the company could also transition to 96-bit technology, with future support for EPC Gen 2 standard.

Operation

At the labeling station, the operator scans the product's bar code and confirms the product run. Operating under their software, the host system interfaces directly with the existing Oracle system to generate the EPC data based on the SKU and production run size. Then the host system receives the EPC data table directly from the Oracle download.

As the products on the conveyor exit the stacker and stretch wrapper at a rate of 23 products per minute, a photo eye detects its presence and sends the signal back to the host system to download the variable EPC data to the SLPA7000. The SLPA verifies the RFID label's integrity, encodes it, and verifies it once again for confirmation—all within a fraction of a second—before printing. Through the same Ethernet connection, the SLPA communicates back to the host system that the RFID label has been applied to the product and if it encountered any quiet/bad RFID labels. After applying the RFID label, the product is scanned by the bar code and RFID scanner. If either scanner cannot read, the product is diverted for rework.

Within approximately 30 seconds, from the start of the process, the products enter a blast freezer at about 40°F below zero for several hours. Prior to storage in a -20°F freezer, the products are palletized and stretch wrapped and enter a manual labeling area where an RFID scanner reads the smart labels on the cases to identify the pallet and downloads the EPC data to another host system interface. Once an operator confirms the pallet is ready to be manually labeled, the host system uploads the EPC data to the Oracle system. The Oracle system generates a unique EPC number that links all the case EPC numbers on the pallet and sends the data back to the host system, which transfers the variable EPC data to a Printronix SL5000e™ MP, located in a warm booth. The SL5000e MP prints out the pallet label and the operator applies it to the pallet. In some line locations, the company has already automated this process by using an SLPA7000 to encode, print and apply the label to the pallet.

Tandem Efficiency

While the pilot experienced a lower than expected incidence of bad and quiet smart labels on the roll, the company quickly realized the cost of rework. With each bad label incidence, the risk increased that product would go unlabeled. In an automated operation with live EPC data, correcting the information requires immediate attention by diverting the cases and returning them to the lines for re-labeling.

For the most efficient solution, the company decided on a zero-downtime tandem system, in which two SLPA7000s are mounted on a single stand and connected via Ethernet to the host system. See Figure 13.2. A sensor located at the stacker and stretch wrapper's exit signals the host terminal that a twin-pack is entering the labeling station. The host system downloads the EPC data via Ethernet to the first SLPA and awaits confirmation that a smart label is printed and dispensed on the tamp pad. If the first SLPA confirms that the RFID label is ready, the host system enables the first SLPA to apply the RFID label. If the host system does not receive confirmation from the first SLPA as a result of a series of bad labels, or if the printer is out of RFID labels, then the second SLPA encodes and applies the RFID label. This dual-machine format increases line efficiency by assuring that every product receives an RFID label, and by providing the ability to change RFID labels instantaneously without affecting production throughput. When the production lines run products that do not require smart labels, the operators simply change out the SLPA's media and the units serve as production bar code applicators

FIGURE 13.2

Tandem print, encode and apply process for dairy products company.

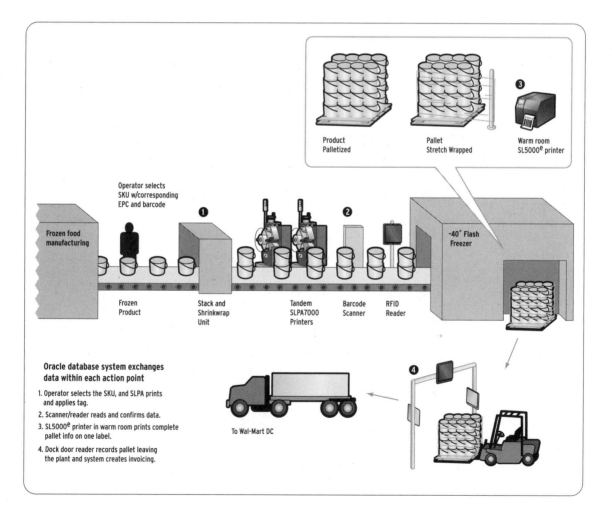

for the non-RFID manufacturing process, allowing the company to retire their older print and apply units.

Conclusions

After thorough research of processes and RFID printing solutions, and incorporating their existing infrastructure into the solution, the company not only met Wal-Mart's RFID mandate but also reaped the benefits of real-time management and greater visibility and communication along the supply chain. Key benefits included:

Increased line efficiency – Products are always bar coded and RFID-enabled for EPC compliant customers which increase product visibility for tracking and inventory control and management.

Increased accuracy – Eliminates data errors with dual redundancy (bar code and EPC), identifies discrepancies in a real-time manner, and provides product uniqueness in the event of product recalls.

Increased reliability – Reduced points of failure with their infrastructure, eliminating the need for middleware and its maintenance.

Case 3: A New Generation Supply Chain for Snack Foods

As one of the world's largest food and beverage companies, this multi-national corporation manufactures, distributes and markets global brands in 200 countries. The company has progressively expanded its reach, achieving sales of greater than $25 billion. As technology development throughout the organization supports its continued growth, the integration of RFID to distribution will enable inventory monitoring and pricing management at the store level, while meeting the requirements of the Wal-Mart mandate.

Currently, the non-RFID distribution process includes identifying, counting, and palletizing pre-printed cases. The pallets are automatically labeled with unique preprinted bar code labels on adjacent sides. The forklift operator scans the license plate number (LPN) and the warehouse management system (WMS) links the palletized products to the customer order. Finally, the pallet is shipped to the designated retailer.

RFID Challenge

To incorporate RFID into the supply chain by the January 2005 Wal-Mart deadline and ultimately attain the maximum operating efficiency from an RFID process, the company sought enterprise alliance partners with real-world experience who could help develop a pilot program by engineering the most flexible, integrated RFID solution. This solution would enable the company and Wal-Mart to document and track the products from the manufacturer's warehouse to the retailer's distribution centers, as well as throughout the supply chain.

Researching RFID solution providers, the company identified key goals for the project:

1. Scalability – to build a repeatable solution that would be engineered to provide growth for the next couple years as the strategy of RFID and retailer expansion unfolded. Scalability also addresses ROI process optimization, while avoiding business disruption by developing a solution in a test area and then expanding the process for broader application.

2. Flexibility – the solution should adapt to different existing manufacturing and distribution practices of the company's product families and would need to accommodate future changes in supply chain management. It should also allow the company to implement a standardized solution across the businesses.

3. Adaptability – required for hardware and software. Multi-protocol EPC Class 0, 0+, and 1 Protocol support provides asset management of the RFID infrastructure, and also allows the company's divisions to select the protocol suitable for their unique requirements since the optimum RFID tag depends upon the product content.

Solution Partners

The company tapped three partners with extensive RFID expertise to meet the goals.

A global leader in supply chain WMS technology solutions would meet the scalability requirement. With its RFID middleware technology, this experienced solution provider interfaced with the existing legacy system and built a warehouse management system. The middleware manages the data by storing all the selected SKUs, and generates, assigns and confirms unique EPC numbers for individual SKU cartons.

A local integrator known for providing complete RFID solutions brought the turnkey flexibility of its RFID solution and data management to work with the middleware. With its bundled solution covering hardware, software, and services, the integrator conducted extensive evaluation with RFID readers and smart label designs and placement in relation to multiple products, pallet patterns, and label durability. This research proved invaluable reducing the manufacturer's learning curve without impacting production or raising costs, and enhancing reliability and RFID performance, The company expects to use this data when it transitions to EPC Gen 2's increased read/write performance.

Printronix was the printer partner chosen for the flexibility and adaptability offered by its smart label printer applicator, the SLPA7000™, co-developed with FoxIV, an Export, PA-based printer applicator manufacturer. The SLPA features RFID Smart™ encoding technology, the foundation for the SL5000ᵉ MP™, the industry's first

RFID-enabled high-frequency, multi-protocol thermal transfer printer, also deployed in this application. The SLPA introduces the industry's first integrated tag management system built into a printer applicator. The technology allows bad and quiet smart labels to remain on the liner, while other systems tested offered limited solutions such as applying defective smart labels to the product or using a mechanical reject arm which allows the smart label to fall to the floor, causing a potential safety hazard. The company also selected the 64-bit Alien Class 1 squiggle tag in 4" x 2" label format to encode the EPC data and print a bar code with human readable text. The printing solution will allow smooth transition to 96-bit technology and EPC Gen 2.

Solution System

The RFID pilot was implemented at three different distribution warehouse locations for two different product lines – convenience foods at one location, and beverage at the other two. Two of the distribution centers used packaged WMS software systems. The third differed significantly, serving as a climate-controlled manufacturing site and distribution center using its own "home grown" WMS system to interface with the middleware. The integrator developed a robust process in a modular RFID printing solution that will ultimately integrate into the warehouses across divisions.

Step One

Beginning with the middleware, pallets are selected for smart labeling. See Figure 13.3. At the labeling station, an operator scans and manually confirms the palletized cartons from a LPN label. The

FIGURE 13.3

Case labeling operation for snack foods shipped to Wal-Mart.

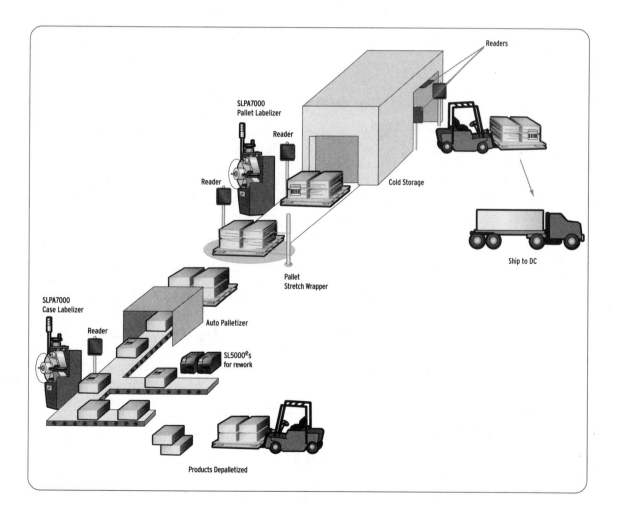

middleware receives confirmation and initializes the system with the unique EPC data for the palletized cartons, and downloads it to the integrator's RFID solution package. As the cartons are manually depalletized and placed on a conveyor, the automated system performs a carton count and downloads the unique EPC data to the SLPA to apply the smart label to the carton. After labeling, the cartons pass an RFID scanner for verification. If the scanner does not detect a smart label, the carton is recycled back to the SLPA.

Step Two

At the end of a pallet run, the system compares the induction count with the ending count and the downloaded EPC data with the scanned EPC data to identify any discrepancies. It documents and alerts the operator of any missing EPC data. The operator confirms the discrepancy as a result of either product damage or a bad or missing smart label. With a bad or missing smart label, the operator will either reload the product onto the induction conveyor or confirm possession of the product at a manual label station. There, the system downloads the missing EPC data to the Printronix SL5000^e MP that prints out the carton label for manual application. The operator loads the carton back on to the conveyor to complete the pallet load. The system counts and documents the EPC data to construct a virtual pallet profile. The cases are then automatically re-palletized and transferred to another conveyor for stretch wrapping.

Step Three

After stretch wrapping, the pallet passes another reader that compares data against the expected profile and advances to a labeling station. The system uploads the EPC data back to the middleware, which generates a unique EPC number that links the EPC products on the pallet. At the manufacturing facility location, the pallet labeling process is completely automated, allowing the pallet to proceed to storage without human intervention. At the other two sites, the middleware then downloads the EPC number back to the system and the manual labeling station. The SL5000e MP prints the smart label and the operator manually applies it to the pallet. The pallets then proceed to storage, pending a Wal-Mart order. Once a load order is received, a forklift operator loads the pallet onto a Wal-Mart truck for delivery to the retailer's designated distribution center.

Conclusions

By conducting a thorough search and selecting enterprise alliance partners, the snack and beverage manufacturer met Wal-Mart's RFID mandate and also reaped targeted benefits.

1. Scalability – Products are always bar coded and RFID-enabled for EPC-compliant customers. The company's solution will allow for growth and expansion.

2. Flexibility – While leveraging both existing and emerging Auto-ID infrastructure by simultaneously generating bar codes and an EPC, the solution can adapt to different product groups to accommodate supply chain changes.

3. Adaptability – The software and hardware infrastructure provided asset protection that allowed the company to use the protocols its applications required.

ELECTRONICS INDUSTRY

Case 4: Smart Partnering for a High-tech Titan

This multi-billion dollar consumer electronics manufacturer is testing and verifying the value RFID technology will bring to retailers, manufacturers, and ultimately to customers. For years the company has experimented with RFID, but early in 2003 it embarked on its RFID compliance adventure by generating its own study on the benefits of RFID implementation in the supply chain, then created a smart partnering strategy to get there.

The company had developed an expertise in tag types, read ranges, tag placement, and automating RFID into production and distribution processes. It has built up an RFID practice among its customers in the manufacturing arena, supporting deployment to comply with the Wal-Mart mandate of January 2005.

Partners for a Pilot

In deploying its own RFID systems in the retail supply chain, the company started a pilot program in January, 2004. Turning to other RFID experts, it established dynamic partnerships in a go-to-market strategy combining robust software with end-to-end services.

Working closely with several middleware providers to integrate RFID into its existing system, the company set the groundwork. For RFID label printing, the project selected Printronix Smart Label Printer Applicators to tag cases and Smart Label Desktop Printers for pallets. Though initially only a few of the many consumer products the company ships to Wal-Mart were EPC-tagged, the entire range of consumer products sold in the stores will soon be readied for tagging.

RFID Challenge

Various contract manufacturers package the company's pilot products at several plant locations. Given the wide variety of products produced, a facility may have products with a one-to-one relationship to the package, or products with a many-to-one relationship. If the product has a one-to-one relationship, the tag is placed on the outside of the packaging. If it has a many-to-one relationship, the tags are placed on the outside of the master cartons. Both situations require a manual application of RFID labels to each pallet, but the faster moving lines require automated encode, print-and-apply process for applying RFID labels to cases.

Solution

Together, the middleware and Printronix RFID hardware address EPC and RFID requirements to tag both individual pallets and high-volume, high-speed conveyors of cases, before being shipped to select distribution centers.

The company needed a solution that could accommodate reliable encoding of tags by Rafsec, Omron and Alien. Early in the pilot, the company utilized the Printronix SL5000e MP RFID printer for printing RFID labels on cases. When Printronix expanded its SmartLine printer family with the SLPA7000 in June 2004, the company conducted extensive review and demonstrations to outline their automation requirements with its capabilities. The company selected the SLPA7000 for its ability to encode, verify, print and apply multiple sizes and EPC classes of RFID labels, and its competence in preventing application of bad tags to cases. A set-up interval occurs between production orders to accommodate case size changes, product size variations, and the changing application heights and distances of the SLPA7000 resulting ad rates. The product's shipping pallets are manually labeled with pallet tags encoded and printed with the SL5000e MP. The Printronix printer uses a multi-protocol UHF encoder (compatible with Class 0, 0+ and 1 tags) to encode multiple tag classes and sizes.

Extensive testing was done on label sizes, RFID tag read rates, and tag types. The exact location and size of the label was critical because of the existing package graphics and geometry of the single item case. Labels from Rafsec, Omron and Alien were eventually selected given the existing shipping label size requirements and read ranges. The original Printronix RFID printers the company installed did not initially support all selected tags, but due to the flexible and modular design of the SL5000^e, the company easily updated their printers with the tag suppliers' support via an update.

Operation

In multiple production lines, flat, collapsed, cartons are automatically placed on a conveyor, formed into shape by a case erector, and then filled with product. By experimenting with the tag placement on the carton, the company learned that if a gap existed between the product and the place on the carton where the tag was attached, they could avoid many interference problems. Even so, work continues in order to achieve 100 percent read rates with the cases on a shipping pallet. The full case is then packaged by taping the case shut, and a bar code is printed on the box. Like other Wal-Mart compliance suppliers, the company will continue to use bar codes in addition to RFID tags. Bar codes are used to identify products as they move through the supply chain, as well as at the retail level, and act as a human-readable backup, should it be required.

The cases activate a box sensor attached to the conveyor, and trigger an "event" in the SLPA7000's General Purpose Input/Output (GPIO) to apply an encoded label. See Figure 13.4. The GPIO is a Printronix Input/Output device manager, designed to enhance the ability to send and receive signals to and from a materials handling system or external device, to complete complex tasks such as system alerts, driving multiple applicators. As the SLPA7000 receives a signal, it advances a

FIGURE 13.4

Contract labeling operation for an electronics manufacturer.

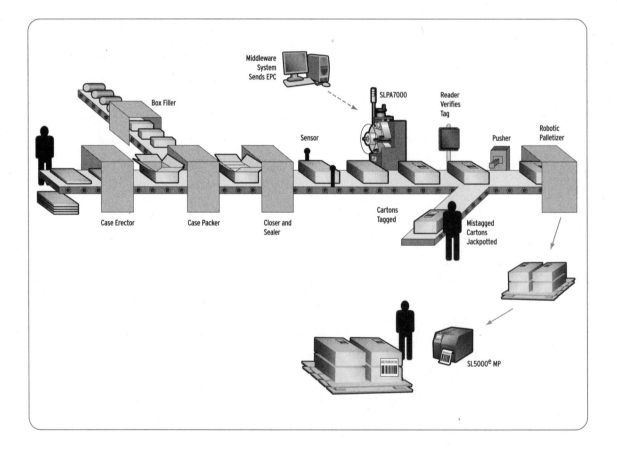

96-bit Class 1, RFID label to be verified, encoded, printed and applied to cases as they move down the conveyor.

Conveyor speeds are approximately 20 boxes per minute. If the SLPA7000 detects a bad or "quiet tag" as the label advances off the roll, an attempt is made to encode several times over the course of a fraction of a second. The bad tag is then over-struck for easy identification, and advanced on the rewind wheel for analysis. This feature provides users traceability of bad tags. It also eliminates the user intervention commonly found in less sophisticated mechanical arm reject designs.

Tagged cases move down the conveyor past an external reader for re-verification of label application, signal strength and read-range of that label. If the case does not have an RFID tag, the case is rejected and diverted off the conveyor into a reject bin for rework. Cases will subsequently be sent down the conveyor a second time for re-application of tags.

Cases with verified, (good) labels move down the conveyor to an accumulation conveyor. Here the cases are stacked on pallets; stretch wrapped, and transferred to a new conveyor. At this point the SL5000e MP desktop RFID printer prints a batch of Class 1, 4" x 6" labels with EPC numbers. These labels are applied to the finished pallets to complete the labeling process. Pallets are moved to dock doors for final verification. An Advanced Shipping Notice (ASN) will be sent to the Wal-Mart distribution center.

When the tagged pallets and cases arrive at Wal-Mart's distribution center, readers at the dock doors automatically scan the tags. The data passes to an application that will alert the company, along with the retailer's operations and merchandising teams that the specific shipment has arrived. Cases are removed from the pallet and processed as usual, then trucked to the participating Wal-Mart stores.

When tagged cases arrive at the back of these stores, the tags on the cases will be read and automatically confirm the arrival of the specific shipment. Signs featuring the EPCglobal logo are placed at the shelf where the company's products are sold to help customers identify tagged items. Understanding and appreciating consumer privacy concerns, the RFID tags do not contain nor collect any additional information about consumers, and do not track customers after the product leaves the store.

Conclusions

In May, 2004, the company began shipping EPC-tagged products to Wal-Mart's Dallas/Fort Worth distribution center as part of the retailer's trial. As the implementation expands, the company sees RFID's potential to drive significant gains in productivity. The company concluded that a pallet of one of its products ready for shipment could be processed in just 11 seconds – down from the 90 seconds it had taken previously. It's understood that in electronics manufacturing there is a high degree of variability, where every item could mean a shift in configuration. But the company realizes products can be shipped and received more quickly, making RFID development worthy of pursuit.

During 2005, several more SLPA7000 printer applicators and SL5000e MP printers will be installed. The company continues to share its growing knowledge of RFID to its business divisions and manufacturing partners.

MILITARY SUPPLIER

Case 5: Transportation Supplier Rides the RF Wave

A 30,000 employee maker of turbines for aircrafts, ocean vessels and railroad engines saw the value of RFID as an information enabler across much of its operation. Given the state of the technology, and the available knowledge in the market, however, the company started small. It knew that at the very least it would fall under a US Defense Department mandate. Beginning with a goal to meet the mandate requirements, the company was confident it could acquire practical knowledge and practices that would help drive a return on the investment.

The company established a leadership team that included those overseeing manufacturing productivity. Areas of responsibility include the DoD liaison, tool tracking, WIP tracking, gage calibration, logistics and vehicle tracking. It also has a marketing liaison, responsible for determining which products were viable candidates for RFID tacking in both the government and commerical sectors of its business. The marketing group also assists a great deal in the feasibility study and business case analysis.

Using a phased approach the company began piloting RFID in the following areas:

Outbound materiel shipments – Smart labels embedded with Alien tags are produced by Printronix printer/encoders. Labels are placed on pallets at the product packaging facility before being loaded for shipment.

Advanced shipping notice generation – Using the Pack Shop System (PSS), tag information is appended to the ASN form for transmission.

Receiving at distribution center – Using Alien readers, parts received at the distribution center are checked in and verified using a Patni Computer System.

Rollout of RFID

Following the successful pilot, the company is advancing its implementation to the following areas:

DoD mandate compliance – Passive UHF tagging of all cases and pallets shipped to depots under DoD guidelines, initially as a voluntary participant.

WIP tracking – Passive tags on totes and subassemblies are used to track work progress through checkpoints.

Kit pulling system – Subassemblies stored in the distribution center in kit form have passive tags. RFID reads are used to trigger events.

WIP platform tracking – Active tags are used on large assembly platforms to track their locations and provide visibility to bottlenecks.

Test hardware tracking – Using active tags and readers on forklifts and buildings, the company is able to track test equipment mounted on engine dollies.

Fleet vehicle tracking – Active tags are used to locate lift trucks and security vehicles.

Torque wrench control – Tags are used to serialize and identify torque wrenches and associated calibration data under industry guidelines.

Gage calibration – Tools that require calibration are tagged for quick identification during calibration.

Conclusions

Drawing on its experiences during the initial stages of its RFID system, the company is now developing best practices to aid in going forward. It now has a good understanding of the competencies needed to conduct a feasibility study and implement an specific RFID application. The company found that much of this work can be done in house, using its experienced engineering and IT staff.

In the area of vendor evaluation and device procurement, the company has a methodology for defining roles, responsibilities and expectations. It only orders what it needs and asks its suppliers to participate in a proof of concept before full procurement takes place. What it did not anticipate in the initial pilot was the downstream costs of an RFID implementation. These costs include those for edge servers, maintenance, support and software to integrate data with their manufacturing execution system. The company now involves its IT staff in assessing the needs of the particular applications, translating those needs to a functionality specification for the software, and evolving standards for operation. The system integraton task tends to be the most lengthy and costly piece, and it largely falls to the internal IT staff to make it work. IT staff also installed the network connectivity and power connections for readers and antennas at each point. In some cases, portals required fixtures that required special fabrication. Again, these resources needed to be identified and time allowed to get the work done.

Although some of its experiences are clearly due to the immaturity of the technology, the company now knows what to watch out for with RFID. For example, it has learned how much harsh use a passive RFID tag and smart label can take. In some instances it places duplicate tags on a pallet to ensure reads.

Finally, the company has learned that getting a payback from RFID can be a challenge, especially since it has always operated as a low volume lean manufacturing operation. It is clear that the upfront investment in education the RFID team and operators will pay off, but continuous communications are necessary to meet its efficiency and customer service goals.

The company has announced they are able to show an ROI through proof of delivery. RFID will provide traceability to reduce lost or damaged goods claims between buyers and suppliers. They believe 80 percent of shipping receipts claims can be reduced or eliminated using RFID as a tracking tool – resulting in substantial cost savings, and increased customer satisfaction.

Sources and Further Reference

RFID in a Box®, see www.manh.com.

Alien RFID tags, see www.alientechnology.com.

Oracle Warehouse Management, see www.oracle.com.

Rafsec RFID tags, see www.rafsec.com.

Omron RFID tags, see www.omron.com.

Patni Computer System, see www.patni.com.

PepsiCo, Inc., see www.pepsi.com.

Products Section

SMART LABEL DEVELOPER'S KIT   352

SMART LABEL PRINTERS   354

ENCODE, PRINT AND APPLY RFID SMART LABEL APPLICATOR SYSTEM   358

RFID SMART LABELS   362

THERMAL BAR CODE PRINTERS – RFID READY   364

PrintNet ENTERPRISE PRINTER MANAGEMENT   368

OPTIONS AND ACCESSORIES   372

GENERAL PURPOSE INPUT OUTPUT MODULE   376

PROFESSIONAL SERVICES   378

www.printronix.com

SMART LABEL DEVELOPER'S KIT

The Printronix Smart Label Developer's Kit is specifically designed to help you fast track an RFID pilot. The kit creates a complete smart label developer's environment, something that cannot be done with an "off-the-shelf" printer.

Transition from Bar Codes to Smart Labels

Printronix's unique RFID Software Migration Tools will allow you to effortlessly move from the printing of bar code labels to encoding smart labels, putting your pilot program on fast track. The tools include a suite of applications that convert standard UPC and GTIN print streams into printer commands, which encode UPC or GTIN on the smart labels, in addition to printing the standard bar code label. Simply select from one of our many migration tools through the printer control panel. The printer will automatically enable the encoding of smart labels with-

out the need to change your print stream, application or environment.

Encoding EPC Compliant Labels

When you are ready to advance your RFID pilot to the encoding of EPC labels, the Developers Kit will grow with you. Our Software Migration Tools include an EPC tool built into the printer, allowing you to use any existing label design software to create an EPC RFID tag. The Printronix Smart Label Developer's Kit also comes with a complete set of programming manuals. Use the step-by-step instructions to add EPC commands to your existing PGL® or other printer data streams.

The kit contains everything necessary to begin the encoding of smart labels:

- SL5000^r MP multi-protocol 4" thermal printer. A web-enabled industrial-grade thermal bar code printer designed for exacting label applications.

- Integrated RFID UHF encoder.

- 1,000 certified RFID smart labels.

- One 625m premium wax ribbon.

- PrintNet® Enterprise Ethernet card and connectivity software.

- Software Migration Tools (SMT™) that permit the seamless encoding of smart labels.

The kit also includes PrintNet® Enterprise, the web-enabled remote network print management system that provides instantaneous visibility to every network printer. It allows users to simultaneously configure and efficiently manage an unlimited number of Printronix printers. This edition also supports management of RFID encoder capabilities.

SMART LABEL PRINTERS

As the adoption of RFID technology accelerates in retail and government supply chain operations, so must the printing solutions for smart labels. SmartLine™ RFID printers enable encoding and printing of various label sizes and antenna designs that have emerged as popular standards through early adopter pilot programs. In addition to label size and antenna design, the need for specific encoding technology to support today's EPC standards with an easy migration path to tomorrow continues to be important.

SL5000^r™ RFID PRINTER

Based on the 5^r Multi-Technology Platform, the SL5000^r is the next generation family of RFID printers. The printers deliver leadership by design, with RFID solutions that address a wide range of applications requiring EPCglobal Class 0, 0+, 1, Gen 2 and Philips 1.19 standards and global frequency requirements. The printers are software upgradeable to support the latest innovations and standards, providing true asset protection.

LEADERSHIP BY DESIGN

- Multi-protocol capabilities support EPCglobal Class 0, 0+, 1, Gen 2 and Philips 1.19 standards and global frequency requirements.

- Intelligent media detection proactively identifies media type to avoid wasting RFID labels with a standard bar code print job.

- PowerPC controller and new high-performance 5^r driver deliver industry-leading throughput.

- EPCglobal Class 1 Gen 2 compatible upgradeability via free firmware upgrade

- Intelligent tag management that identifies and overstrikes substandard smart labels.

ADDITIONAL FEATURES

- Integrated RFID UHF encoder meets global frequency requirements.

- Smart™ Label encoding validation and overstrike capability.

- PGL, ZPL™ and SATO RFID programming language support.

- Unique dual motor ribbon system eliminates clutch replacement and ribbon wrinkle.

- Snap-in print head allows operator to easily replace print heads and change from 203 dpi to 300 dpi printing without firmware or software changes.

- Wireless/Ethernet option provides real-time data access and local printing flexibility.

- Online Data Validation option offers a complete bar code/RFID compliant solution.

- Printronix PrintNet® Enterprise option provides instantaneous visibility to every network printer and allows simultaneous configuration and management of an unlimited number of Printronix printers.

- Greater ribbon capacity eliminates frequent attention and disruption to workflow.

ASSET PROTECTION

The Printronix SL5000^r smart label printers provide RFID solutions compliant to today's standards and compatible for growth to expand to tomorrow's needs. Built on the 5^r Multi-Technology Platform, the SL5000^r offers a foundation for protocol upgradeability such as expansion to EPCglobal capabilities via firmware upgrades, including future upgrades to allow EPCglobal Class 1 Gen 2 compatibility. To offer global compliance, these printers can be switched to other regional frequencies via a global upgrade kit.

SMARTLINE PRINTER SPECIFICATIONS

SL5000ʳ RFID PRINTERS

SL5000ʳ MP	Multi-protocol UHF encoder set to global frequency standards
	Supports for Class 0, 0+, 1,
	Philips 1.19 standards*
	*Compatibility with future EPC global Class 1 Gen 2 standard through firmware upgrade. Please contact Printronix or a Printronix Certified RFID Integrator for availability.

PRINTING CHARACTERISTICS

Print Speed	SL5204ʳ: 10 IPS @ 203DPI (254mm/sec)
	SL5304ʳ: 8 IPS @ 300DPI (203mm/sec)
Printing Methods	Thermal Transfer or Direct Thermal
Resolution	203/300 DPI (operator interchangeable)
Printable Width	4.1" (104mm) max

RFID ENCODING MODES

Integrated RFID reader and antenna assembly

Operation Modes	Write/Verify/Print – writes RFID data to tag and verifies contents are written correctly, while also printing the desired image
Error Handling Modes	Overstrike – when a bad RFID tag is detected, Overstrikes label and applies the data with the next label
	Stop – when a bad tag is detected, stops the printer to allow for user invention
Statistics Tracking	Tracks number of tags written to and number of bad tags detected

MEDIA HANDLING CHARACTERISTICS

Tear-Off	Individual label tear-off
Tear-Off Strip	Label strips tear-off
Continuous	Labels print continuously
Peel	Labels peel from liner without assistance (peel mode requires rewind option)

MEDIA HANDLING OPTIONS

Rewinder	Required for peel and present. Not recommended for batch rewind of RFID labels
Cutter Kit	Cuts labels after printing specified number of labels

MEDIA COMPATIBILITY

Media Types	Roll or fanfold
	Labels, tags and tickets
	Paper, film or synthetic stock
	Thermal Transfer or Direct Thermal
Media Width	1.0" to 4.5" (SL5204ʳ/SL5304ʳ)
Media Thickness	0.0025" to 0.010"
Roll Core Diameter	3.0" (7.6 cm)
Maximum Roll Diameter	8.0" (20.9 cm)
Thermal Transfer Ribbon	
Ribbon Width Range	1.0" to 4.33" (SL5204ʳ/SL5304ʳ)
Ribbon Capacity	625m

OPERATOR CONTROLS & INDICATORS

Operator Controls	Off Line-On Line, Test Print, Job Select, Form Feed Menu, Cancel, Enter
Message Display	32 character
Indicators	Off Line-On Line, Menu

BAR CODE VALIDATION

Optional	Online Data Validation (ODV) – verifies bar code quality, overstrikes failed bar codes, and a replacement label is printed

PROGRAMMING LANGUAGES

Standard	Printronix Graphics Language (PGL)
	Zebra Graphics Language (ZGL)*
	TEC Graphics Language (TGL)*
	Intermec Graphics Language (IGL)*
	Sato Graphics Language (STGL)*
	*Printer Protocol Interpreters for ZPL, TEC, IPL
	and Sato with RFID commands for ZPL and
	Sato only

PROTOCOLS

Optional	Telnet TN5250/TN3270

BAR CODE SYMBOLOGIES AVAILABLE

AUSTPORT, Aztec, BC35, BC412, CODABAR, Code 11, Code 35, Code 39, Code 93, Code 128 (A, B, C) DATAMATRIX, EAN8, EAN13, FIM, 125GERMAN, Interleaved 2 of 5, ITF14, Matrix, MAXICODE, MSI, PDF417, PLANET, PLESSEY, POSTNET, POSTBAR, ROYALBAR, RSS14, TELEPEN, UCC/EAN-128, UPC-A, UPC-E, UPC-EO, UPCSHIP, UPS11

SENSING METHODS

Transmissive, Reflective (Gap, Mark, Notch, Continuous Sensing Form)

INTERFACES

Standard	Serial RS232
	IEEE 1284 (Centronics)
Optional	Ethernet (includes PrintNet Enterprise
	Software CD)
	Wireless (802.11b) (includes PrintNet
	Enterprise Software CD)
	Coax/Twinax
	GPIO (General Purpose Input/Output)

FONTS, GRAPHICS SUPPORT, WINDOWS DRIVERS

Fonts	OCRA, OCRB, Courier, Letter Gothic CG
	Triumvirate Bold Condensed
Graphic Support	PCX and TIFF file formats
Windows Drivers	Windows NP/2000/XP

MEMORY

DRAM	32Mb standard
Flash	8Mb standard (16Mb optional)

POWER REQUIREMENTS

Line Input	90–264 VAC (48–62Hz), PFC
Power Consumption	150 watts (typical)
Regulatory Compliance	FCC-B, UL, CSA, ETSI EN 300 220, CE

ENVIRONMENTAL CONSIDERATIONS

Operating Temperature	5°C to 40°C
Dimensions	11.7" W x 20.5" L x 13.0" H
Printer Weight/Shipping Weight	38 lbs/47 lbs

ENCODE, PRINT AND APPLY RFID SMART LABEL APPLICATOR SYSTEM

The SLPA7000™ is a unique encode, print and apply RFID Smart Label applicator, co-developed by Printronix and FOX IV Technologies. Built on Printronix leading EPCglobal RFID Smart encoding technology and combined with FOX IV's best-in-class Uniwall applicator system, the SLPA7000 is the first and only all-in-one system to meet RFID compliance encoding and encode and apply manufacturing production requirements. Combining both RFID smart label printing technology with applicator capability, this RFID smart label solution delivers fast, accurate, cost-effective encoding, printing and application to users with site specific requirements.

FULLY INTEGRATED SYSTEM

- Multi-protocol encode, print and apply capabilities.

- Intelligent tag management identifies and overstrikes substandard smart labels.

- Supports EPCglobal Class 0, 0+ and Philips 1.19 standards.

- FCC-B compliant solution.

- EPCglobal Class 1 Gen 2 compatible upgradeability via free firmware upgrade.

ADDITIONAL FEATURES

- Only encode-print-apply system to prove interoperability at the EPCglobal Interoperability Test, compatible with multiple RFID technology suppliers.

- Performance levels ranging from 20-40 labels per minute.

- Identifies and rejects "bad" or "quiet labels" – insuring 100 percent high-performance RFID labels.

- Top and side label application orientations, as well as roll-on/front/back.

- Precise label placements within 1/16" on stalled carton or 1/4" @ 100 ft/min.

- Fully integrated controller board, reducing the costs and complexity found in a traditional Programmable Logic Controller (PLC).

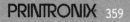
ASSET PROTECTION

The SLPA7000 is expandable and upgradeable –
providing broader long-term capabilities for your
supply chain requirements, reducing the risk of
obsolescence. Built on the SL5000^e MP multi-
protocol platform, the SLPA7000 offers a foundation for protocol upgradeability and expansion through firmware updates such as expansion to EPCglobal capabilities, including future
upgrades to allow EPCglobal Class 1 Gen 2
compatibility.

SMARTLINE PRINT, ENCODE AND APPLY SPECIFICATIONS

SLPA7000 SMART LABEL PRINTER APPLICATOR

SLPA7000	Multi-protocol support for Class 0, 0+, 1, Philips 1.19
	915MHz UHF encoder

PRINTING CHARACTERISTICS

Print Speed	10 IPS* @ 203DPI (254mm/sec)
	8 IPS* @ 300DPI (203mm/sec)
	*the system throughput may vary due to bad tag processing
Printing Methods	Thermal Transfer or Direct Thermal
Resolution	203/300 DPI (operator interchangeable)
Printable Width	4.1" max

RFID ENCODING MODES

Integrated RFID reader and antenna assembly

Supports Class 0, 0+, Philips 1.19 tags

Operation Modes	Write/Verify/Print – writes RFID data to tag and verifies contents are written correctly, while also printing the desired image
Error Handling Modes	Overstrike – when a bad RFID tag is detected, overstrikes the label and prints and encodes the data onto the next label and then applies
Statistics Tracking	Tracks number of tags written to and number of bad tags

MEDIA COMPATIBILITY

Max Roll Size	3" ID x 12" OD
Max Label Length	10"
Min Label Width	1" (25mm)
Max Media Width	4.5" (112mm)
Method of Detection	Label gap, hole, back mark

OPERATOR CONTROLS & INDICATORS

Operator Controls	Off Line-On Line, Test Print, Job Select, Form Feed Menu, Cancel, Enter
Message Display	32 character
Indicators	Off Line-On Line, Menu

PROGRAMMING LANGUAGES

Standard	Printronix Graphics Language (PGL)
	Zebra Graphics Language (ZGL)
	TEC Graphics Language (TGL)*
	Intermec Graphics Language (IGL)*
	*Printer Protocol Interpreters for ZPL, TEC, and IPL with RFID commands for ZPL only

BAR CODE SYMBOLOGIES AVAILABLE

CODABAR, Code 39, Code 93, Code 128 (A, B, C) DATAMATRIX, EAN8, EAN13, FIM, Interleaved 2/5, ITF14, MAXICODE, PDF417, POSTNET, POST-BAR, ROYALBAR, RSS14, UCC/EAN-128, UPC-A, UPC-E, UPC-EO, UPS11

SENSING METHODS

Sensing Methods	Transmissive, Reflective (Gap, Mark, Notch, Continuous Sensing Form)
Ribbon	625m length
	4.33" width
	Model 8500 premium wax formulation
Media Width	1.0" to 4.5"

INTERFACES

Standard	Serial RS232
	IEEE 1284 (Centronics)
Optional	Ethernet (PrintNet)
	Dual Ethernet/wireless (802.11b)

FONTS, GRAPHICS SUPPORT, WINDOWS DRIVERS

Fonts	OCRA, OCRB, Courier, Letter Gothic CG
	Triumvirate Bold Condensed
Graphic Support	PCX and TIFF file formats
Windows Drivers	Windows 95/98/2000/XP

MEMORY

DRAM	8Mb standard (16Mb optional)
Flash	4Mb standard (10Mb optional)

UTILITIES/ENVIRONMENTAL REQUIREMENTS

Line Input	90–264 VAC (48–62Hz), PFC
Power Consumption	150 watts (typical)
Regulatory Compliance	FCC-B
Compressed Air	80 to 100 PSI, 1/4" NPT connector
Operating Temperature	5°C to 40°C
Humidity	Max 85 percent RH, noncondensing
Dimensions	29.0625" L x 17" W x 24.5" H
Weight	96 lbs (43.6 kg)

RFID HANDLING

Frequency	902-928 MHz
	Encoder upgradeable

PRINTER ORIENTATION

360° rotatable display accommodates common printer orientations: nose down, nose up, bottom apply, top apply and horizontal

RFID SMART LABELS

Printronix SmartLine RFID printers support the widest selection of RFID tags in the industry. This allows you to use a single platform to support the various applications and requirements for RFID labeling in your logistics supply chain. Printronix has adopted a collaborative, non-proprietary strategy to openly encourage the availability of reliable RFID labels for Printronix RFID printers. To support this strategy we are continuing to release Guide Specifications to the RFID community for specifying RFID labels for use in Printronix RFID printers.

These specifications outline the requirements for RFID labels and placement of RFID inlays for best perfomance in the Printronix family of RFID printers. Specifications are available for downloading from www.printronix.com.

Certified Printronix Smart Labels

An abridged list of certified Printronix RFID Smart Labels are listed in the following table. For a complete list, see www.printronix.com.

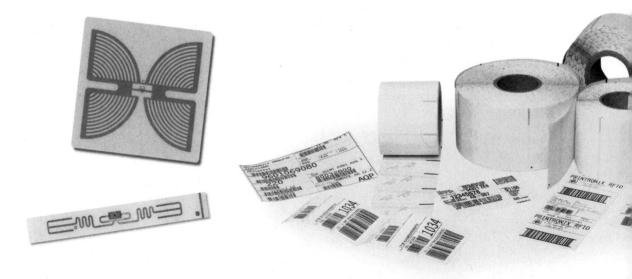

EPC Class 1, UHF 902 to 928 MHz (915 MHz Nominal)

Inlay	EPC Memory	Label Sizes	# of Labels per roll	Printronix Ordering Number
Alien Squiggle	96 bits (read/write)	4" x 2"	1,000	178933-001
		4" x 4"	750	178933-002
		4" x 6"	500	178933-003
		4" x 8"	400	178933-004
		4" x 2"	3,000 fanfold	178933-201
		4" x 6"	1,000 fanfold	178933-203
		4" x 2"	3,400	178933-401
		4" x 4"	1,800	178933-402
		4" x 6"	1,200	178933-403
		4" x 2"	500	178933-405
		4" x 6"	500	178933-406

UCode 1.19, UHF 853 to 883 MHz (869.525 MHz Nominal)

Inlay	EPC Memory	Label Sizes	# of Labels per roll	Printronix Ordering Number
Rafsec 450	96 bits (read/write)	2.8" x 2.8"	1,000	178891-001
KSW PH58	96 bits (read/write)	4" x 4"	750	Call for info
		4" x 6"	1,000	Call for info

THERMAL BAR CODE PRINTERS − RFID READY

T5000^r™ thermal printers, based on the world-class 5^r Multi-Technology Platform, are web-enabled, industrial-grade thermal bar code printers designed to operate in demanding manufacturing or distribution environments. Through leadership by design, Printronix thermal printers enable the rollout of bar code label printing throughout the supply chain with an open migration path to RFID for future requirements. Users can obtain RFID upgrade kits as needed when supply chain applications require advance tracking and compliance. Upgrade kits have everything needed to upgrade to a multi-protocol SmartLine™ printer.

Optimized with PrintNet® Enterprise, the T5000^r printers allow IT network managers to remotely control, configure, manage and monitor printers over the network. And with Online Data Validation (ODV™) and ODV Data Manager, users have a closed loop system providing the security of 100 percent guaranteed good bar codes with documented readability reports.

LEADERSHIP BY DESIGN

- Software upgradeable assures support of the latest features.

- Widest range of emulations for easy consolidation of multiple printer brands.

- PowerPC controller and new high-performance 5^r driver deliver industry-leading throughput.

- Snap-in print head allows operator to easily replace print heads and change from 203 dpi to 300 dpi printing.

- RFID field upgradeable supports future RFID requirements.

ADDITIONAL FEATURES

- Unique dual motor ribbon system eliminates clutch replacement and ribbon wrinkle.

- Aluminum die-cast design dampens vibration and maintains precise printer alignment.

- Ventless system operates in environments with airborne particulate matter without compromising performance.

- Easy side-load, cleaning and maintenance.

- Wireless/Ethernet option provides real-time data access and local printing flexibility.

- The GPIO option offers unparalleled integration with third party equipment.

- Greater ribbon capacity eliminates frequent attention and disruption to workflow.

- ODV option offers a complete bar code/RFID compliant solution.

- PrintNet Enterprise option provides total control of worldwide print operations.

ASSET PROTECTION

The Printronix T5000^r thermal printers provide an upgrade path to RFID through a comprehensive upgrade kit solution compliant to today's standards and compatible for growth to expand to tomorrow's needs. Built on the 5^r Multi-Technology Platform, the T5000^r offers a foundation for protocol upgradeability such as expansion to EPCglobal capabilities via firmware upgrades. Available upgrades include global frequency kits and RFID upgrade kits for multi-protocol EPCglobal Class 0, 0+, 1, Philips 1.19 and future EPCglobal Class 1 Gen 2. The T5000^r eliminates service calls with field installable snap-in print heads that allow easy resolution change and the most extensive list of options such as ODV, guaranteeing 100 percent good bar codes every time.

THERMALINE PRINTER SPECIFICATIONS

T5000^r RFID PRINTERS

T5000^r	Thermal transfer or direct transfer
	Industrial grade
	RFID field upgradeable

FIELD UPGRADE KIT

Upgrade your thermal printer to RFID with the following upgrade kit

SLMP Kit	Multi-protocol UHF encoder set to global frequency standards
	Supports for Class 0, 0+, 1, Gen 2, Philips 1.19 standards*
	*Compatibility with future EPC global Class 1 Gen 2 standard through firmware upgrade. Please contact Printronix or a Printronix Certified RFID Integrator for availability.

(Excludes 8" printers T5208^r/T5308^r)

PRINTING CHARACTERISTICS

Print Speed	T5204^r-4": 10 IPS @ 203DPI (254mm/sec)
	T5304^r-4": 8 IPS @ 300DPI (203mm/sec)
	T5306^r-6": 10 IPS @ 203DPI (254mm/sec)
	T5306^r-6": 8 IPS @ 300DPI (203mm/sec)
	T5308^r-8": 8 IPS @ 300DPI (203mm/sec)
	T5308^r-8": 6 IPS @ 300DPI (152mm/sec)
Printing Methods	Thermal Transfer or Direct Thermal
Resolution	203/300 DPI (operator interchangeable)
Printable Width	4.1" (104mm) max (T5204^r/T5304^r)
	6.6" (168mm) max (T5206^r/T5306^r)
	8.5" (216mm) max (T5208^r/T5308^r)

RFID ENCODING (OPTIONAL UPGRADE KIT)

(Excludes 8" printers T5208^r/T5308^r)

UHF encoder set to global frequency standards

Supports EPCglobal Class 0, 0+, 1, Gen 2 and Philips 1.19 standards

Operation Modes	Write/Verify/Print – writes RFID data to tag and verifies contents are written correctly, while also printing the desired image

Error Handling Modes	Overstrike – when a bad RFID tag is detected, Overstrikes label and applies the data with the next label
	Stop – when a bad tag is detected, stops the printer to allow for user invention
Statistics Tracking	Tracks number of tags written to and number of bad tags detected

MEDIA HANDLING CHARACTERISTICS

Tear-Off	Individual label tear-off
Tear-Off Strip	Label strips tear-off
Continuous	Labels print continuously
Peel	Labels peel from liner without assistance (peel mode requires rewind option)

MEDIA HANDLING OPTIONS

Rewinder	Required for peel and present. Not recommended for batch rewind of RFID labels
Cutter Kit	Cuts labels after printing specified number of labels

MEDIA COMPATIBILITY

Media Types	Roll or fanfold
	Labels, tags and tickets
	Paper, film or synthetic stock
	Thermal Transfer or Direct Thermal
Media Width	1.0" to 4.5" (T5204^r/T5304^r)
	2.0" to 6.8" (T5206^r/T5306^r)
	3.0" to 8.75" (T5208^r/T5304^r)
Media Thickness	0.0025" to 0.010"
Roll Core Diameter	3.0" (7.6 cm)
Maximum Roll Diameter	8.0" (20.9 cm)
Thermal Transfer Ribbon	
-Ribbon Width Acceptable	1.0" to 4.33" (T5204^r/T5304^r)
	2.0" to 6.8" (T5204^r/T5304^r)
	3.0" to 8.75" (T5204^r/T5304^r)
-Standard Ribbon Length	625m

OPERATOR CONTROLS & INDICATORS

Operator Controls	Off Line-On Line, Test Print, Job Select, Form Feed Menu, Cancel, Enter
Message Display	32 character
Indicators	Off Line-On Line, Menu

BAR CODE VALIDATION

Optional	Online Data Validation (ODV) – verifies bar code quality, overstrikes failed bar codes, and a replacement label is printed

PROGRAMMING LANGUAGES

Standard	Printronix Graphics Language (PGL)
	Zebra Graphics Language (ZGL)*
	TEC Graphics Language (TGL)*
	Intermec Graphics Language (IGL)*
	Sato Graphics Language (STGL)*
	*Printer Protocol Interpreters for ZPL, TEC, IPL and Sato with RFID commands for ZPL and Sato only

PROTOCOLS

Optional	Telnet TN5250/TN3270

BAR CODE SYMBOLOGIES AVAILABLE

AUSTPORT, Aztec, BC35, BC412, CODABAR, Code 11, Code 35, Code 39, Code 93, Code 128 (A, B, C) DATAMATRIX, EAN8, EAN13, FIM, 125GERMAN, Interleaved 2 of 5, ITF14, Matrix, MAXICODE, MSI, PDF417, PLANET, PLESSEY, POSTNET, POSTBAR, ROYALBAR, RSS14, TELEPEN, UCC/EAN-128, UPC-A, UPC-E, UPC-EO, UPCSHIP, UPS11

SENSING METHODS

Transmissive, Reflective (Gap, Mark, Notch, Continuous Sensing Form)

INTERFACES

Standard	Serial RS232
	IEEE 1284 (Centronics)
	USB 2.0
Optional	Ethernet (includes PrintNet Enterprise Software CD)
	Wireless (802.11b) (includes PrintNet Enterprise Software CD)
	Coax/Twinax
	GPIO (General Purpose Input/Output)

FONTS, GRAPHICS SUPPORT, WINDOWS DRIVERS

Fonts	OCRA, OCRB, Courier, Letter Gothic CG
	Triumvirate Bold Condensed
Graphic Support	PCX and TIFF file formats
Windows Drivers	Windows NP/2000/XP

MEMORY

DRAM	32Mb standard
Flash	8Mb standard (16Mb optional)

POWER REQUIREMENTS

Line Input	90–264 VAC (48–62Hz), PFC
Power Consumption	150 watts (typical)
Regulatory Compliance	FCC-B, UL, CSA, ETSI EN 300 220, CE

ENVIRONMENTAL CONSIDERATIONS

Operating Temperature	5°C to 40°C
Dimensions	11.7" W x 20.5" L x 13.0" H (T5204^r/T5304^r)
	13.4" W x 20.5" L x 13.0" H (T5204^r/T5304^r)
	15.4" W x 20.5" L x 13.0" H (T5204^r/T5304^r)
Printer Weight/	38 lbs/47 lbs (T5204^r/T5304^r)
Shipping Weight	40 lbs/49 lbs (T5204^r/T5304^r)
	43 lbs/52lbs (T5204^r/T5304^r)

PrintNet ENTERPRISE PRINTER MANAGEMENT

PrintNet® Enterprise is the only system that lets you manage all your print operations—whether it's a single warehouse or a global infrastructure —from one computer. This advanced Web-based management tool combines a fully integrated Ethernet adapter and Java-based remote management software to deliver unparalleled remote printer management adaptability, remote diagnostics and help desk tools. It supports SNMP, which allows you to use SNMP managers like HP Open View, Tivoli, Computer Associates Unicenter TNG, Sun Net Manager and Castle Rock SNMPc.

With PrintNet Enterprise you get:

- Instant RFID printer status, which indicates which printers are in use or idle and which need consumables or repairs.

- Instant alerts, which notify you via e-mail (or routed to your pager or cell phone) if an RFID printer situation requires immediate action.

- Instant control, which allows you to remotely change printer settings, update firmware, download fonts and lockout local operators from making changes.

- Instant RFID printer organization, which enables you to configure printers from a single action.

- Instant diagnostics, which remotely diagnoses problems via a Web browser.

Visibility

Whatever the status of your printers, you will receive a message with a clear and comprehensive view of your entire printer operation.

- **Main Status Screen** – Displays the status of every printer under management, including a "gas gauge" indicating ribbon levels and color-coded icons highlighting problem spots and text messages.

- **SNMP** – Fully supports the printer MIB, making PrintNet® Enterprise-equipped printers visible and accessible with tools such as HP OpenView, WebJet Admin, Tivoli, Computer Associates Unicenter TNG, Sun Net Manager, and Castle Rock SNMPc.

Instant Alerts

You are instantly notified when a printer situation demands immediate attention. Whether by e-mail, pager or cell phone, these instant messages can be generated to a specific responder, varying according to class of alert. Even if the printer is totally dead due to catastrophic power supply, controller failure, or loss of power, an instant alert is generated.

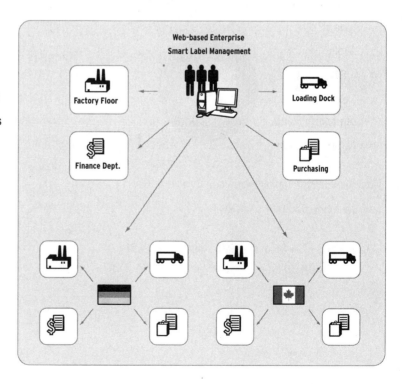

Web-based Enterprise Smart Label Management

Factory Floor

Loading Dock

Finance Dept.

Purchasing

Alerts can be delivered in numerous ways

- Email Message
- SNMP Traps
- Syslog
- Logged to File

Remote Presence

From any network PC or workstation, you can quickly diagnose a problem without the need to travel to the printer.

- **Web Access** – Configuration management and remote control panel are accessible over the web using standard web browsers.

- **Remote Control Panel** – View and operate any printer's control panel online as if it is local.

- **Remote Configuration Access** – View and modify printer configuration settings.

- **Remote Flash File Access** – View, download, and upload user files in the printer.

Whatever the situation demands, you have total control of your printers. Change printer settings, update firmware, download fonts, clear buffers – even lock out local operators from making changes. All these controls are available to you without ever going to the printer.

- **Configuration Editor** – Creates, saves, and downloads configuration tables consisting of all parameters.

- **Configuration Locking** – Controls which printer settings, firmware updates and downloaded fonts can and cannot be modified by local operators, including the ability to completely exclude them from making any changes.

- **Flash File Manager** – Add, remove and move files (such as fonts and predefined forms) stored in the printer's flash file area.

- **Automated Firmware Updates** – Firmware versions of one or more printers can be updated at the same time. All configurations and files are automatically retrieved and restored as part of the update.

- **Speed Keys** – One-touch access to change the most commonly accessed printer settings.

- **Automatic Configuration Switching** – Tie a printer configuration to one of eight internal print queues. Whenever a job is sent to that queue, the specified configuration is automatically loaded. This feature ensures that your jobs will be processed with the correct configuration every time.

Centralized Programming

You can ensure all printers have a common configuration by programming once and sending the programming to all your printers.

- **Individual, Group or Broadcast Management** – Perform an operation to entire groups of printers with one set of commands. There's absolutely no need to configure printers individually.

- **Printer Database** – Allows organization of printers into groups matching applications, locations, or any other criterion the environment demands.

OPTIONS AND ACCESSORIES

RFID Mobile Print System

Today's just-in-time (JIT) manufacturing and logistics environments require advanced printers that can perform where and when they are needed. Although wireless LAN technology provides a viable solution, it also increases the need for true printer management to ensure consistency and diagnostic control.

Printronix SmartLine Mobile Print System (MPS) addresses this need. Combined with our RFID printer, MPS delivers smart labels directly to locations that need it the most. Its ergonomic design, durable steel frame and heavy-duty wheels combine maximum maneuverability with minimal maintenance to offer these benefits:

- Reduced centralized printing and manual distribution by printing smart labels on the spot.

- Scalable deployment for maximum printing flexibility at compliance locations.

- Elimination of expensive, time-consuming, hard-cabled system reconfigurations when changeovers occur.

- Centralized control, consistency and diagnostics previously available only in hard-cabled managed systems.

- Optimized cost efficiency and productivity.

- Higher return on investment (ROI) due to accelerated data and inventory movement.

Once installed in a wireless network, and powered by PrintNet Enterprise, this unique technology allows printing of real-time data in any location. It eliminates time-consuming and expensive re-configuration of hard-cabled systems when warehouse or manufacturing floor changeovers occur. It also provides centralized control, consistency and diagnostics previously available only in hard-cabled PrintNet Enterprise-managed systems.

Integration into Legacy and Proprietary Language Environments

Printronix provides tools that enable the T5000^e, SL5000^r and SLPA7000 to integrate seamlessly into virtually any proprietary language environment. This enables greater bar code labeling efficiency, and ultimately, a link to higher performance and reliability in the delivery cycle of your products. Printronix Printer Interpreter (PPI) integrates ThermaLine and SmartLine printers in Zebra, Intermec and TEC installations easily. Once integrated, they maintain all capabilities used in a Printronix environment.

TN5250/3270 Ethernet Option

As the market moves toward 100 Base-T and 1 Gigabit Ethernet, maintaining an out-dated Twinax or Coax network becomes increasingly difficult to justify. Two power factors impair a migration: legacy applications must be changed and a high level of print job control must be maintained. The TN5250/3270 Ethernet Option maintains the job control and visibility of Twinax or Coax networks, yet combines it with the sped and efficiency of Ethernet. In addition, the option provides:

- Automatic confirmation of which pages have and have not printed.
- Knowledge of exactly where an error has occurred in batch printing.
- Ability to restart printing on only those pages needed.
- Full access to all the features and functionality of PrintNet Enterprise.

On-Line Data Validation (ODV™)

Printronix Online Data Validation (ODV) is an option exclusive to our T5000^e thermal printer and SmartLine RFID series. It's a tool that can save time, money, stalled productivity... and a lot of frustration. This unique quality control option can inspect every bar code on every label printed on a Printronix T5000^e or SL5000^r. Specifically, ODV technology analyzes each bar code just after it is printed. This 100 percent analysis verifies that the print image falls well within each bar

code's published quality specifications to ensure the bar code will scan successfully. If any code on a label fails, the label is automatically cancelled and a replacement label is printed.

Today's supply chains depend on Automatic Identification and Data Capture (AIDC) technology for proper routing and distribution. AIDC allows no tolerance for bar code scanning failure while products pass through the supply chain. Bar codes must read 100 percent of the time. Any reading failure can escalate into a rejected item, manual recovery steps, time delay, a loss in productivity, an unhappy customer, compliance failure fines, or a even a loss of business.

Printronix created Online Data Validation to help you avoid such problems. In doing so, ODV:

- Evaluates every bar code on every label to provide total scanning assurance.
- Automatically cancels and reprints any label on which a bar code failed.
- Eliminates the need for costly human intervention in the bar code validation process.

• Combined with Printronix PrintNet® Enterprise, ODV closes the loop on bar code quality control and provides unprecedented visibility of mission-critical print operations.

ODV Data Manager

Imagine a world in which bar code verification systems could integrate with enterprise networks, eliminating costly bar code rejection and capturing real-time data about printing applications. Imagine if customers could monitor and review all bar codes printed and then merge the data into custom reports, all while improving Return on Investment. With our exclusive ODV Data Manager integrated with PrintNet Enterprise, imagination is not needed. ODV Data Manager provides robust data capturing and exporting capabilities and the ability to evaluate data within each bar code. It merges information into databases and ports it to any application, such as SAP or Oracle.

ODV Data Manager helps customers manage industrial network printer systems, while providing a safeguard against the high costs associated with bar code scanning and data accuracy failure.

The Printronix Online Data Manager provides the following functions:

• Ability to capture bar code and validation data.
• Ability to export captured data to either an ASCII file or SQL database.
• Ability to stop the device and alert the operator of a problem.

All data captured by the ODV Data Manager is date and time stamped, and associated with the printer generating the data. Collected data is either stored in a history file or immediately exported to an external SQL database. Data stored in the history file can be saved in either a comma separate variable (CSV) file format, or as an XML report. Data from a single printer or multiple printers can be saved in either a single file or in multiple files.

GENERAL PURPOSE INPUT OUTPUT MODULE

Printronix has expanded its growing portfolio of printer management solutions and enhanced the functionality of its thermal printers with the General-Purpose I/O (GPIO) Module and GPIO Manager. The GPIO accessory module is designed specifically for those who want to integrate Printronix T5000^e thermal printers with other systems, such as label application systems, programmable logic controllers, light stacks and host computers.

The GPIO accessory module is comprised of an I/O circuit board that is easily installed in a T5000^e printer. Simple printer menus allow programming of specific interface signals for proper polarity or logic functions that can meet all typical interface print/apply requirements or be compatible with the functions of all major competitors' interfaces.

When combined with the GPIO Manager software, the GPIO module leverages Printronix's exclusive solutions enablers, such as the Online Data Validation (ODV™) system, resulting in the industry's most cost-effective solution, using the least amount of hardware.

The GPIO Manager software is a flexible, PC-based program offering simple and intuitive programming tools, including a graphical user interface (GUI) with pull-down menus, "self-documenting" with an audit trail for each line, and the ability to edit and upgrade quickly on a single screen. Additionally, the GPIO Manager software can program and map desired actions from the T5000^e series printer and GPIO accessory module to pre-defined events, such as "printer online", "label present", "label taken" and "printer error".

The GPIO accessory module contains eight optically isolated inputs, eight optically isolated outputs and four relay outputs, all available for GPIO Manager mapping. In addition, printer functions, such as keypad switches, serial communication ports and proprietary ODV analysis parameters, are available with the GPIO Manager, extending its mapping functions.

Integrators of the T5000e series printers with label application systems can program a seamless system with simultaneous functions when a label is printed. For example, signals are sent to the label applicator, a light stack is illuminated for visual bar code or printer status

indications, and printing resumes after the label applicator has received the label. Additionally, printer keys can be programmed with special functions for custom user access, and specially formatted priority data transmissions can be sent to a host for specific conditions. The system allows virtually boundless creative and economic solutions for value-add opportunities.

PROFESSIONAL SERVICES

Your environment can be simple or complex. Your printers can be connected directly to your host, or scattered on networks around the globe. Your production line might have verifiers validating printed bar code labels as they move down a conveyor. No matter how your environment is set up, leveraging the full capability of your printers and verifiers may require consultation, development, customization, or training to ensure maximum productivity.

Printronix Professional Services can deliver what you need.

Many companies require their trading partners to adhere to strict guidelines and standards as integral components of the automated supply chain, or else face the prospects of returned goods or even fines. Printronix has the proven experience to handle retailer and vendor supply chain requirements, freeing your organization to focus on its core business.

It's not just about sending print jobs. It's about global printer management – error detection, notification and correction; remote troubleshooting; remote software updates; and more. It's not just about verifying bar codes on a label. It's about how your operation can run non-stop with full assurance that critical steps in your process, which depend on successful bar code scanning, will take place without fail.

Auto-ID/RFID Consulting

As a member of EPCglobal, Printronix Professional Services can provide a full range of expertise, from bar coding to RFID. Our specialists can provide on-site assessments, training and integration of bar code and RFID systems. We can guide you through a rapid implementation of smart label deployment, and assist in developing and implementing a successful migration strategy from bar code to RFID.

Closed-Loop Bar Code Verification

Printronix Professional Services can provide customers a full auditing of every label and bar code printed. Our specialists can provide integration to external corporate databases, including Oracle, DB2, SQL/Server, MySQL and others. We can develop custom management reports to address compliance, charge-backs and track quality control metrics, and create custom software to identify erroneous bar code content such as duplicate or out of sequence serial numbers.

Label Compliance and Certification

Using our expertise and close association with industry leaders, we can assess root cause for failed labels and bar codes against any automotive, retail or pharmaceutical specification. Printronix Professional Services is a GM and Sears approved certification provider. We can provide assessment and solutions for nearly all label related charge back fines. We canimplement a complete auditing and quality control system to record label and bar code quality history, and can create compliance labels to meet any specification.

SAP and ERP Bar Code Integration and Implementation

Printronix Professional Services can facilitate the integration of SAP/ERP printing through custom device types and drivers. We offer ABAP/4 and SAP script support and are experts in all available middleware packages.

Legacy Application Migration and Implementation

We can migrate data streams to new printing platforms without modifying host applications. We offer services to redesign and rewrite applications for new environments. We can apply an array of software tools to bridge the translation, as well as provide Microsoft Back Office integration.

For the location of your nearest Printronix representative, call 800-826-3874.

Printronix, Inc.
14600 Myford Rd.
P.O. Box 19559
Irvine, CA 92623-9559

www.printronix.com

SINGAPORE
ASIA REGIONAL SALES OFFICE
Printronix Schweiz GmbH
No. 42, Changi South Street 1,
Changi South Industrial Estate
Singapore 486763
Tel: (65) 6542-0110
Fax: (65) 6546-1588

AUSTRALIA
AUSTRALIA/NEW ZEALAND OFFICE
Printronix Australia Pty. Ltd
Level 21, 201 Miller Street,
North Sydney NSW 2060
Tel: (61-2) 99592250
Fax: (61-2) 99592244

CHINA
SHENZHEN REGIONAL SALES OFFICE
Printronix Printer (Shenzhen) Co. Ltd
Unit F, 17/F Shenzhen International Trade
Commercial Building Nan Hu Road,
Luo Hu District, Shenzhen China 518014
Tel: (86-755) 25194027
Fax: (86-755) 25194019

BEIJING SALES OFFICE
Printronix Beijing
Room 1831, 18F, China Merchants Tower
No 118 Jian Guo Road
Chao Yang District
Beijing 100022 PR China
Tel: (86-10) 65662731
Fax: (86-10) 65662730

SHANGHAI OFFICE
Printronix Shanghai
31F, Jin Mao Tower,
88 Shi Ji Avenue
Pudong Shanghai
200120 PR China
Tel: (86-21) 2890 9719
Fax: (86-21) 2890 9219

INDIA
ASIA PTE REGIONAL SALES OFFICE
Printronix Asia Pte Ltd
B202 2nd Floor, Crystal Plaza,
Link Road, Andheri (W)
Mumbai 400053, India
Tel: (91-22) 26733423
Fax: (91-22) 26733422

KOREA
Printronix Asia Pte Ltd Korea Liaison Office
#3001, 30th Floor, Trade Tower,
159-1, Samseong-dong, Gangnam-gu
Seoul 135-729, Korea
Tel: (82-2) 60072066
Fax: (82-2) 60072723

EUROPE
Printronix Europe, Middle East and Africa
Sales and Marketing Headquarters:
PRINTRONIX FRANCE Sarl
8, rue Parmentier
F-92800 Puteaux
France
Tel: +33 (0) 1 46 25 19 00
Fax: +33 (0) 1 46 24 19 19
Email: EMEAsales@printronix.com

Index

3PL, 9, 13, 282

A.T. Kearney, 4, 24, 100, 140, 154, 159, 275

Absorption, XII, 158, 217, 246

Active tag, XII, 37, 40, 165, 347

Adaptive Frequency Agility, XII, 31

Advanced shipping notice (ASN), XII, 118, 134, 143, 144, 194, 343, 346

Afghanistan, 165

AIM-USA, XII,

Air interface, XII, 42, 47, 59, 62, 73, 77, 100, 218

Albertsons, 2, 9, 146

Alien Technology, 36, 58, 82, 136, 191, 224, 275

Amplitude Modulation (AM), XII, 29

Anti-collision, XII, 59, 257

Anti-theft system, XII

APEC, 87

Application Level Events (ALE), XII, 9, 261, 263, 270, 275

Attenuation, XIII, 64, 66

Australia, 9, 31, 153

Auto-ID Center, XIII, 41, 71, 85, 101, 261, 268

Auto-ID Labs, 86, 151, 218

Autonomous mode, 60, 257

Avery Dennison, 36

Backscatter, XIII, 33, 38, 58, 222

Beam power tag, XIII

Bioterrorism Act, XIII, 10, 153, 160, 254

Boeing Company, 167

British Telecom, 148

Bureau of Customs and Border Protection, XIV, 9, 154

CAGE, XIII, 92, 175

Canada, XV, 9, 29, 87, 147, 153

Carrefour, 2, 9, 147, 149

Case analysis, 54, 116, 119, 189, 196, 208, 212, 228, 243

CEN, 87

Central Asia, 163, 165

Certified label, XIII, 120

Chain of custody, 10, 116, 134, 269

Charge-back, 118, 128, 229

China, 151, 153, 159, 380

Circular polarization, XIII, 66, 216

Code 39, 84, 173

Collision avoidance, XIII, 51

CompTia, 204, 206

Conductivity, XIII

Cover-coding, 77, 104

CPG, XIV, 9, 291, 318

C-TPAT, XIV, 9, 10, 155, 254

Curtain, XIV, 64, 66, 130, 142, 242

Cyclic redundancy check (CRC), XIV, 48, 76, 104, 105

Data cleansing, 272, 273

Database, X, XX, 19, 59, 61, 85, 97, 101, 105, 116, 133, 200, 236, 253, 257, 265

DC (distribution center), XIV, 57, 139, 146, 160, 186, 237, 323, 329, 335

Dead zone, XIV, 223, 226

Decibel, XIV

Defense Logistics Agency (DLA), XIV, 162, 165, 177

Dell Computer, 7

Dense reader, XIV, 76, 78, 80, 259

Department of Defense (DoD), XIV, 2, 6, 9, 11, 23, 42, 70, 88, 91, 100, 110, 130, 138, 163, 165, 177, 180, 185, 199, 346

De-tuning, 68

DFARS, 166, 169

DHL, 12

Dielectric, XIV, 221, 223

Diffraction, 224

Diorio, Dr. Chris, 70, 73, 75

Dipole, XIV, 36, 114, 212

DN-Systems Enterprise Solutions GmbH, 101

Dover, MD Air Force Base, 168

E3, 173

EAN, XV, 86, 90, 188, 357

EAS, XV

ECCnet, XV

EDI, 87, 134, 143, 146, 174, 303, 308, 311

EEPROM, XV, 34, 53

Electromagnetic field, XV, 220

Electrostatic discharge (ESD), XV, 115, 119, 155, 188, 223, 243

EM Micro, 42, 148

EMI, XV, 127, 268

EN 300-220, 31, 32

Encoder, XV, 9, 19, 26, 34, 40, 45, 52, 69, 100, 115, 119, 128, 187, 200, 201, 231, 252, 281, 340, 346, 353, 354

Encryption, 51, 99, 102, 267

EPCglobal, XI, XVI, 9, 43, 45, 56, 69, 73, 82, 83, 86, 92, 103, 107, 108, 147, 158, 169, 175, 182, 202, 231, 261, 268, 275, 282, 344, 354, 358, 365, 378

ERP, XVI, 3, 9, 135, 192, 197, 238, 255, 271, 277, 281, 298, 325, 379

Error recovery, 53, 56, 111, 123, 133, 229

Ethernet, XIX, 62, 142, 236, 253, 266, 268, 325, 328, 352

European Telecommunication Standards Institute (ETSI), 30, 56, 87, 357

Far field, XVI, 122

FCC, 30, 32, 56, 59, 66, 87, 357

Federal Express, 11

Field probe, 217

Field strength, XVI, 218

Firmware, XVI, 21, 55, 63, 135, 199, 202, 259, 266, 355, 358

Fixed RFID reader, XVI, 27, 31, 38, 58, 65

GPIO, XVII, 244, 342, 357, 365, 376

GRAI, XVII, 90, 91, 175,

Grocery Manufacturers of America (GMA), 4, 5

Grunwald, Lukas, 101

GS1, XVII, 86, 87

GTAG, XVII

GTIN, XVII, 55, 90, 91, 133, 294, 352

Gulf War, 164

Half-duplex, XVII, 59

Hall, James D., 164

Handheld RFID reader, XVII, 46, 57, 58, 78, 145, 199, 262, 266, 307

HAZMAT, 111

Healthcare Distribution Management Association, 158

HERF/HERO/HERP, 173

Hertz, XVII, 28

High frequency (HF) tag, XVII, 30, 42, 157, 158, 160, 215

Highjump Systems Inc. (3M), 289, 316

Host computer, XVIII, 26, 27, 38, 41, 54, 60, 61, 76, 108, 111, 124, 127, 131, 235, 236, 245, 250, 253, 257, 259, 264, 268, 325, 328, 377, 378

IBM, 3, 24

IC, XVII, 33, 34, 49, 81

Impinj, Inc., 36, 42, 70, 82, 108

Inductive coupling, XVII, 122

Inlay, XVII, 26, 32, 34, 112, 114, 119, 129, 363

Integral RFID, 218, 226

Intellident, 148

Interference, XVII, 29, 40, 48, 51, 59, 68, 76, 80, 127, 184, 187, 208, 211, 213, 217, 223, 224, 341

International Paper, 138

Interoperability testing, 69, 82, 174

Interrogator, XVII

InTransit Visibility Network (ITV), 165, 166

Iraq war, 165

Iron mountains, 164

ISM bands, XVIII, 29, 44

ISO, 9, 43, 45, 48, 71, 80, 86, 87, 88, 102, 103, 152, 243

Japan, 30, 31, 48, 80, 87, 152, 153

Johnson & Johnson, 138, 158

Just-in-case logistics, 164

Just-in-Time (JIT), 40, 163, 372

Kill command, XVII, 42, 49, 61, 74, 78, 102, 103, 149

Kimberly-Clark, 187

Korea, 31, 152, 380

Kraft, 138

KSW Microtec, 36

Label converter, 114, 115, 119, 128, 243

Labor management, 307

LAN, 235, 253, 267, 372

Lessons learned, 23, 118, 141, 154, 168, 181, 184, 185, 187, 191, 209, 221, 241, 318-350

License plate, 19, 101, 116, 257, 331

Linear polarization, XVIII, 67, 215, 216

Line-of-sight, XVIII, 92

Liquid, XIII, 116, 158, 171, 211, 220, 221, 226, 246, 319, 320

Load Management, 278, 280, 284

Logistics, XIV, 12, 24, 145, 163, 164, 168, 169, 177, 237, 254, 293, 301, 305, 308, 310, 311, 315, 319, 346, 362, 372

Lowes, 271

Low-frequency tag, XVIII

Mad Cow disease, 8, 9, 153

Manhattan Associates, 305-316, 320

Marks & Spencer, 148, 160

Mass adoption, 14, 39, 84, 158

Metal, XIII, 36, 64, 68, 96, 116, 167, 213, 221, 223, 224, 226, 246, 319, 320

Metro Store, 2, 9, 101, 138, 147-149, 160

Microchip, XVIII, 26, 32, 37

Microwave tag, XVIII

Middle East, 164, 381

Middleware, 111, 134, 186, 193, 261, 262, 270, 313, 324, 330, 333, 335, 339, 342, 379

Mil Std 129P, XVIII, 116, 171, 175, 177

Military air transport system, 163

MIT, XI, 9, 71, 72, 85, 226

MPHPT, 87

Nano-technology, XVIII, 32

Near field, XVIII, 53, 119, 122, 123, 157, 230

Network print management, XVIII, 134

Network traffic, 134, 256, 265

Nokia, 7

NSI, 88

Ocean shipping, 37, 154

Omron, 36, 340, 341, 350

On-demand, X, 54, 110, 233, 234

One forward, one back, 157

On-line data validation (ODV), 126, 356, 365, 374, 375, 377

ONS, XVIII, 269

Opaque, XVIII, 223, 225, 247

Operation Desert Storm, 165

Operation Iraqi Freedom, 165

Oracle, 327, 329, 350, 375

Order cycle time, XIX, 311

Out-of-stock, XIX, 4, 14, 15, 17, 140

Overstrike, XIX, 53, 190, 200, 354, 356

Oxycontin, 158

Package engineering, 228

Palletizing, 198, 218, 231, 233, 237, 322, 331

Passive tag, XIX, 37

Performance based logistics, 164

PET substrate, 32

Pfizer, 158

Pharmaceutical, 106, 116, 140, 156-160, 171, 211, 359

Philips Semiconductor, XVII, 36, 43, 55, 160, 354

PML, XIX, 134, 269, 270

Portal, XIX, 46, 48, 50, 57, 63, 64, 130, 142, 150, 153, 166, 168, 178, 192, 210, 214, 237, 263, 300, 309, 322, 348

POS, XIX

Power over Ethernet (PoE), XIX, 62, 266, 268

Procter & Gamble, 138

Programmable logic controller (PLC), 244, 358, 376

Purdue Pharma, 158

Quality of service, 267

Quiet label, XX, 126, 190, 358

Radio waves, XX, 26, 122, 221

Radiolucent, XX

Rafsec, 36, 42, 340, 350, 363

Ramstein, Germany USAFB, 168

Random number, 49, 77, 78, 104, 267

Read only tag, XX, 40

Read range, XX, 19, 26, 30, 41, 46, 50, 53, 69, 93, 127, 131, 157, 188, 194, 338, 341, 343

Read rate, XX, 46, 47, 51, 63, 68, 79, 95, 115, 131, 142, 168, 189, 191, 208, 210, 212, 219-228, 232, 245, 285, 322, 340, 341

Read write tag, XX, 101

Readability, 37, 46, 49, 66, 94, 95, 105, 114, 116, 118, 127, 172, 208, 218, 283, 364

Real time locating system (RTLS), 307

Redundancy, XX, 50, 76, 98, 99, 104, 217, 330

Replenishment, XX, 143, 146, 148, 194, 214, 315

Retail link extranet, 145

Return on investment (ROI), X, XXI, 12, 17, 100, 184, 200, 205, 246, 255, 278, 281, 308, 324, 332, 349, 373

Reverse logistics, 315, 316

RFID in a Box®, 305, 313, 314, 316, 320, 350

Rockwell, 159, 206

Router, XIX, 62

Rumsfeld, David, US Secretary of Defense, 162

San Joaquin, CA, 166, 170

Sarbanes-Oxley Act (SOX), XXI, 9, 10, 254, 255

Savant, XXI, 261

Schroeter, John, 70

Security, 8, 9, 10, 29, 30, 42, 49, 51, 99-108, 142, 155, 156, 159, 255, 267, 275, 347, 364

Semi-passive tag, XXI, 40

Sense and respond logistics, 163-169

September 11, 2001 World Trade Center bombing, 8

Serial number, 19, 21, 87, 89-91, 96, 175, 282, 290, 297, 314, 379

Serialization, XXI, 14, 98, 157, 270

SFTP, 174

Shanghai, China, 151, 380

Shenzhen, China, 151, 380

Shielding, XXI, 221, 222, 263

Shipping container, XXI, 9, 90, 91, 146, 173

Shrinkage, XXI, 2, 14, 15, 106, 184, 240, 302, 305, 315

Silent treewalking, 267

Simple Network Management Protocol (SNMP), 59, 265, 368

Singulation, XXI, 47, 79, 121, 123

SKU, XXI, 89, 146, 235, 238, 263, 282, 284, 285, 289, 290, 291, 320, 321, 322, 326, 329, 333

Slap & ship, XXII, 182, 186, 231-243, 281, 282

Slotting, 278, 287

Smart Label Developer's Kit, 191, 352

Spy chips, 18

SSCC, XXII, 90, 91, 174

Stanford University, 7, 24, 388

Supply chain execution system, 103, 111, 133, 266, 299, 300, 319

Susquehanna, PA, 166, 170

Symbol Technology, 36, 42, 44, 82, 136

Target, IX, 2, 9, 61, 74, 100, 138, 146, 196, 221

Tesco, 2, 9, 138, 146-148

Texas Instruments, 36, 160

TMS, XXII

Tote, XXII, 16, 96, 128, 130, 148, 347

Trading Partner Management, 305, 310, 316

Translucence, 220

Transportation Management, 193, 277, 299, 305, 311

Trash compactor, 65, 143

Trigger, 46, 55, 75, 105, 133, 167, 194, 218, 242, 253, 257, 259, 260, 264, 265, 267, 302, 303, 314, 342, 347

UC Irvine Extension, 204, 205

UCC, XXII, 86, 90, 132, 188, 250

UID, XXIII, 90, 91, 170-176

Unilever, 138

United Parcel Service, 12

UPC, XXIII, 17, 21, 23, 75, 84-87

US Air Force, 163, 168

VeriSign Inc., 113, 268, 269, 275

Viagra, 158

Wal-Mart, IX, 2, 4, 36, 42, 45, 82, 95, 130, 138-146, 151, 157, 159, 167, 170, 180, 199, 206, 237, 245, 271, 279, 280, 318-325, 330-344

Warehouse Management System (WMS), XXIII, 3, 9, 41, 192, 197, 238, 255, 281, 288-291, 295, 306, 321, 331, 333

Wavelength, XXIII, 28, 36, 122, 157

Wide Area Work Flow, 174

Wi-Fi, 266

XML, XXIII, 201, 269, 375

Yard management, 277, 294-297, 310, 312

About the Authors

Robert A. Kleist, CEO, President and Director   Bob Kleist founded Printronix in 1974 and serves as its president, CEO and a director. His previous experience included co-founding Pertec Computer Corporation and serving in engineering and management assignments at Ampex, Link Aviation and Magnavox. Kleist holds 17 patents for peripheral and control systems, and he received a BSEE degree from Kansas University and an MSEE degree from Stanford University. He has served on the Seagate Technology board of directors, the Stanford Engineering Advisory Committee, and U.C. Irvine Engineering and Information and Computer Science Advisory Committees. He is a co-author of the 2004 book, "RFID Labeling: Smart Labeling Concepts & Applications for the Consumer Packaged Goods Supply Chain."

Theodore A. Chapman, Senior Vice President, Engineering and Product Marketing; Chief Technology Officer   Andy Chapman is the senior vice president of engineering and product marketing and chief technology officer of Printronix. Chapman joined Printronix in November 1995 as vice president, product development. In 1999, he was appointed as senior vice president, engineering and chief technology officer. From 1970 to 1995, Chapman held various engineering and senior management positions with IBM Corp, focused primarily on high performance tape subsystems, mass storage libraries and printing systems. He holds a B.S. Engineering degree from UCLA and an MSEE degree from Stanford University.

David A. Sakai, Vice President, Marketing David Sakai is the vice president of marketing at Printronix Inc. Before joining Printronix, Sakai served as vice president of channel operations for Xerox Corp.'s Office Solutions Group. During his 24-year tenure at Xerox, Sakai held numerous senior management positions throughout the organization including general manager; vice president of marketing, light lens business unit; vice president of marketing, office systems group and vice president, channel operations office systems group. He is a co-author of the 2004 book, "RFID Labeling: Smart Labeling Concepts & Applications for the Consumer Packaged Goods Supply Chain."

Brad S. Jarvis, Director, Product Marketing With nearly 20 years of broad-based sales and marketing experience at companies such at Xerox and Lantronix, Jarvis oversees the product and marketing strategy for all of Printronix's product lines. He was instrumental in driving the positioning, strategy and product road map for Printronix's first foray into the emerging RFID space in September 2003. Jarvis has lectured extensively and has been quoted as a global expert on RFID. He co-authored, "RFID Labeling: Smart Labeling Concepts & Applications for the Consumer Packaged Goods Supply Chain."